DNA Deamination and the Immune System

AID in Health and Disease

Molecular Medicine and Medicinal Chemistry

Published:

MicroRNAs in Development and Cancer
edited by Frank J. Slack (Yale University, USA)

Merkel Cell Carcinoma: A Multidisciplinary Approach
edited by Vernon K. Sondak, Jane L. Messina, Jonathan S. Zager, and Ronald C. DeConti (H Lee Moffitt Cancer Center & Research Institute, USA)

Forthcoming:

Fluorine in Pharmaceutical and Medicinal Chemistry: From Biophysical Aspects to Clinical Applications
edited by Véronique Gouverneur (University of Oxford, UK) and Klaus Müller (F Hoffmann-La Roche AG, Switzerland)

Molecular Exploitation of Apoptosis Pathways in Prostate Cancer
by Natasha Kyprianou (University of Kentucky, USA)

Antibody Drug Discovery
edited by Clive R. Wood (Bayer Schering Pharma, Germany)

Volume 3 Molecular Medicine and Medicinal Chemistry

DNA Deamination and the Immune System

AID in Health and Disease

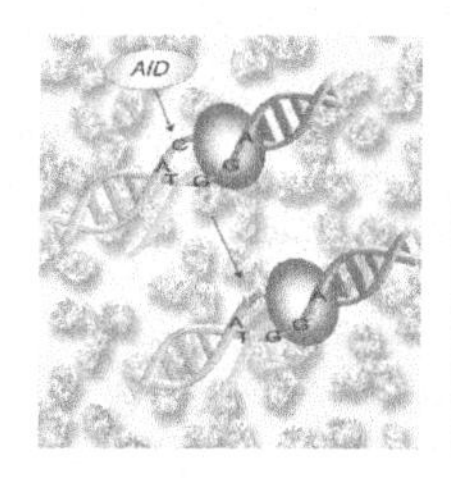

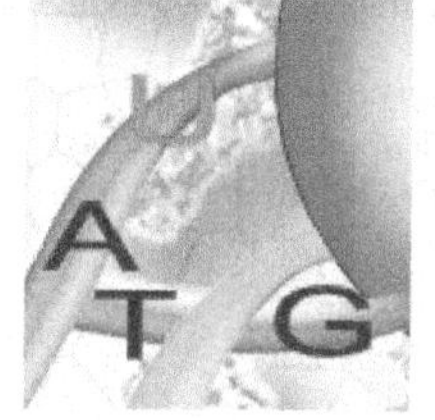

Editors

Sebastian Fugmann
National Institute of Health, USA

Marilyn Diaz
National Institute of Health, USA

Nina Papavasiliou
Rockefeller University, USA

ICP Imperial College Press

Published by

Imperial College Press
57 Shelton Street
Covent Garden
London WC2H 9HE

Distributed by

World Scientific Publishing Co. Pte. Ltd.
5 Toh Tuck Link, Singapore 596224
USA office: 27 Warren Street, Suite 401-402, Hackensack, NJ 07601
UK office: 57 Shelton Street, Covent Garden, London WC2H 9HE

British Library Cataloguing-in-Publication Data
A catalogue record for this book is available from the British Library.

Molecular Medicine and Medicinal Chemistry — Vol. 3
DNA DEAMINATION AND THE IMMUNE SYSTEM
AID in Health and Disease

ISBN-13 978-1-84816-592-2
ISBN-10 1-84816-592-7

Printed in Singapore.

Preface

For decades, studies of the mechanism of somatic hypermutation and class switch recombination had been the focus of only a small group of B cell immunologists world-wide. In the fall of 2000, Prof. Honjo and his coworkers at Kyoto University, Japan, and Dr. Anne Durandy and her colleagues at the Hôpital Necker-Enfants Malades in Paris, France, reported their break-through discovery that the enzyme, activation induced cytidine deaminase (AID), is an essential component of the molecular machinery performing both processes. Since then, the number of scientists around the world working on this intriguing protein and the processes it catalyzes has increased dramatically. Thus, in the fall of 2008, a first mini-symposium that was solely focused on AID was held in Chapel Hill, North Carolina, USA. It sparked a dynamic environment where the most pressing questions in the field were discussed in depth. Over 40 speakers from the laboratories at the forefront of AID research presented their latest exciting findings, and the overwhelmingly positive response to the meeting prompted us to assemble this monograph focused on the findings presented. The nine chapters are co-authored by junior up-and-coming researchers and eminent senior scientists in the field, and provide the reader with a consensus comprehensive overview of our current knowledge about AID itself, the processes it catalyzes, and the burning questions these scientists are trying to address in the future.

Beyond the well characterized role of AID in human hyper IgM syndrome, deregulation of AID has recently been linked to autoimmune disease. AID has also emerged as a major player in the development of B cell lymphomas, as well as a number of other cancers, where it has been correlated both with progression toward malignancy and with relapse. Finally, active DNA demethylation by AID is emerging as a likely non-immune function, with implications both for normal development and tumorigenesis. These recent developments have resulted in a further expansion in the ranks of scientists who are interested in this enzyme. We hope that this monograph will not only serve as a reference point to immunologists, but also to a larger cohort of scientists and physicians including those interested in cancer and stem cell biology.

Sebastian D. Fugmann
Marilyn Diaz
F. Nina Papavasiliou
(Editors)

Contents

List of Tables

List of Figures

Chapter 1

Introduction

F. Nina Papavasiliou[1] *and Janet Stavnezer*[2]

[1]*The Rockefeller University*
New York, NY 10021, USA
E-mail: papavasiliou@rockefeller.edu

[2]*University of Massachusetts Medical School*
Worcester, MA 01655, USA
E-mail: Janet.Stavnezer@umassmed.edu

Biological information is coded in the base sequence of DNA and/or RNA. It follows that the fidelity of this information is meticulously preserved during its replication, transcription and maintenance, particularly in higher organisms; alterations at the level of genome or transcriptome can have dramatic downstream functional implications. Unintended sequence changes often lead to deleterious consequences. However, despite the tremendous pressure to guard against such effects, many biological systems have developed targeted mechanisms which alter DNA or RNA sequences and their corresponding information content. Though such sequence modification pathways have diverse roles throughout biology, many provide important host defense functions in innate and adaptive immunity.

Programmed sequence alterations that change genomic DNA or the genetic meaning of a genomically-encoded transcript have been termed "editing" (Grosjean *et al.*, 2004). To date, the only known enzymatic activities involved in polynucleotide editing catalyze either the deamination of cytidine to uridine in tRNA, mRNA or DNA (Conticello *et al.*, 2007) or the deamination of adenosine to inosine in tRNA or mRNA (Hamilton *et al.*, 2010). This book focuses on the central role of activation-induced cytidine deaminase (AID) in establishing the genetic diversity required for an effective humoral adaptive immune response by editing DNA at immunoglobulin (Ig) loci.

1.1 Discovery of AID

The technical catalyst for the discovery of AID was the generation, in 1996, of the B lymphocyte cell line, CH12F3 (Nakamura *et al.*, 1996). Derived from the CH12.LX lymphoma cell line, CH12F3 was selected to undergo class switch recombination (CSR) at a high frequency and exclusively to the isotype IgA upon stimulation with IL-4, TGFβ and CD40L. Theorizing that a specific recombinase was responsible for CSR, Muramatsu and Honjo (Muramatsu *et al.*, 1999) applied a PCR-based subtraction method to screen genes upregulated upon stimulation of CH12F3 cells for class-switching. Among the four novel genes discovered, AID proved to be especially interesting because of: its germinal center B cell restriction; its homology to the APOBEC family of RNA cytidine deaminases; and its *in vitro* deaminase activity, unique from that of APOBEC–1. Furthermore, AID-deficient mice were unable to undergo CSR and, surprisingly, also somatic hypermutation (SHM; Muramatsu *et al.*, 2000), underscoring the central role of the protein in Ig diversification reactions. Concurrently, through the use of standard human genetics, the Durandy laboratory independently identified AID as the gene responsible for HIGM in a subset of patients with an autosomal form of hyper-IgM syndrome (henceforth named type II, or HIGM2) who had a severe CSR defect and lacked somatically-mutated immunoglobulin genes (Revy *et al.*, 2000).

Initial sequence comparison of AID revealed that it possessed a cytidine deaminase domain with homology to the only well-characterized RNA deaminase at the time, which was a protein termed Apolipoprotein-B mRNA editing catalytic polypeptide–1 (APOBEC–1; Muramatsu *et al.*, 1999). APOBEC–1 was known to specifically edit the mRNA of apolipoprotein-B, converting a glutamine to a stop codon and thereby generating two distinct apoB isoforms with different functions

(Wedekind *et al.*, 2003). Based on this homology, it was initially hypothesized that AID also edited a single mRNA that functioned in both SHM and CSR. Though this was an entirely reasonable proposition at the time, a wealth of additional evidence over the years has clearly tipped the balance toward the notion that AID directly edits DNA at the Ig locus.

1.2 Current Model of AID Function

Experimental support for the hypothesis that AID is a DNA deaminase was first provided by showing that ectopic expression of AID results in mutation of the *E. coli* genome (Petersen-Mahrt *et al.*, 2002). As it is unlikely that AID would edit the same mRNA in prokaryotic and eukaryotic cells to generate a novel DNA mutator, the simplest interpretation of these data is that AID is a *bona fide* DNA mutator, and as such, the first member of a family of polynucleotide deaminases that act on DNA. Other studies demonstrated that ectopic expression of AID was able to mutate the genome of a number of mammalian cell types (plasma cells, HEK–293T cells, NIH 3T3 cells; Martin *et al.*, 2002; Yoshikawa *et al.*, 2002), and also of yeast (Mayorov *et al.*, 2005; Poltoratsky *et al.*, 2004).

Further experimental support for the notion that AID is a DNA mutator emerged from studies focused on an important intermediate – U:G mismatches. Specifically, the Neuberger laboratory conducted experiments to study the role of uracil DNA glycosylase (UNG) in SHM and CSR. UNG is the major glycosylase that removes uracil from DNA in the context of base excision repair. Thus, if AID does indeed catalyze the formation of uracil in DNA, UNG would be expected to be central to the resolution of U:G mismatches. Indeed, Rada and colleagues (Rada *et al.*, 2002) found that UNG-deficient animals mutated their Ig locus at rates identical to those of their wild-type littermates. However, the spectra of mutations that accumulated at Ig sequences were very different. Specifically, mutations at G and C were strongly biased towards C to T and G to A events, as a result of direct replication of such mismatches, and the mutations observed at A and T bases were similar for both *ung*$^{-/-}$ and wild-type mice. In addition to its importance for SHM, Rada and colleagues found that UNG was central to CSR. In the absence of UNG, CSR is nearly abrogated (Rada *et al.*, 2004). The notion that UNG is central to SHM and CSR was further bolstered by experiments in the Durandy laboratory, which found that a subset of HIGM patients carried mutations in their UNG genes (Imai *et al.*, 2003).

Finally, a number of groups have studied the biochemical activities of purified AID from several expression systems (activated B cells, baculovirus-infected insect cells, recombinant expression in bacteria; Bransteitter *et al.*, 2003; Chaudhuri *et al.*, 2003; Dickerson *et al.*, 2003) and shown that AID exhibits activity only on single-stranded DNA substrates (naked DNA; Bransteitter *et al.*, 2003; Dickerson *et al.*, 2003), transcribed double-stranded DNA (Besmer *et al.*, 2006; Chaudhuri *et al.*, 2003) or transcribed DNA complexed with nucleosomes (Shen *et al.*, 2009), but not on non-transcribed double-stranded DNA. In addition, the local sequence preference of AID *in vitro*, namely that the WRC motif is an AID activity hotspot (Pham *et al.*, 2003), coincides well with SHM hotspots observed *in vivo*: changing the AID coding sequence results in concomitant changes in DNA hotspot motif preferences (Wang *et al.* 2010).

1.3 Open Questions

AID has been shown to be an active DNA mutator in a number of settings. It follows that in the cell AID must be meticulously regulated, and this is indeed the case: AID is regulated transcriptionally, post-transcriptionally and post-translationally (Ramiro and Di Noia discuss AID regulation in Chapter 7).

An additional mode of regulation appears to be the targeting of the molecule to the Ig locus, and locus-specific elements important for SHM/CSR targeting have been described (and are discussed by Dunnick and Fugmann in Chapter 3). Chromatin modifications important for SHM/CSR targeting have also been described, as well as the presence of local sequence features (Gearhart and Kenter, Chapter 2).

It is important to distinguish between AID targeting to the Ig loci and repair targeting that eventually results in SHM/CSR. AID could be directly targeted to the Ig loci via a factor or factors that tether it to each locus (discussed by Reina-San-Martin and Chaudhuri in Chapter 4). A plausible non-mutually exclusive alternative is that faulty resolution of uracil lesions is mostly unique to the Ig loci. Krijger *et al.* will discuss error-prone and error-free lesion resolution in SHM (Chapter 6); CSR lesion resolution is the topic for Yu and Lieber (Chapter 5). Experiments in mice (Rada *et al.*, 2004) and zebrafish (Rai *et al.*, 2008) have demonstrated a genetic association between AID and UNG as well as AID and thymine DNA glycosylase (TDG): i.e. whereas UNG is clearly downstream of AID in a pathway that results in error-prone repair at the Ig locus, a genetic association of TDG with AID appears to be required for active CpG demethylation in zebrafish, a process which is error-free. Though there is no evidence thus far that AID directly interacts with either of those molecules, it is intriguing that the AID:UNG pathway results in error-prone repair, in contrast to the AID:TDG pathway for which the associated outcome is thought to be faithful repair of uracil lesions. The intriguing possibility therefore remains that AID itself selects between error-prone and error-free repair, possibly through the acquisition of cell- or locus-specific interaction partners, or cell- or locus-specific post-translational modifications.

Finally, lack of AID clearly causes immune deficiency. Conversely, AID overexpression (or ectopic expression) is also thought to be causal for a number of cancers of many different tissues. It is not hard to imagine that off-target action of AID in the germinal center can cause mutations and eventually even cancer-causing translocations (such as the well-studied *IgH:c-myc*). Indeed, mutations due to AID activity have been documented at a number of oncogenes and tumor-suppressors isolated from germinal center cells (e.g. *p53, c-myc, pim1, bcl-6*: Liu *et al.*, 2008; Pasqualucci *et al.*, 2001; Shen *et al.*, 1998). The link between AID expression and occurrence of B cell lymphoma is strong, and is discussed by Willmann *et al.* (Chapter 8), as well as by Diaz *et al.* (Chapter 9). Curiously, AID has also been detected in a large number of different tumors (e.g. breast, prostate, bone marrow-derived lymphomas, hepatocellular carcinomas, gastric cancers; Chiba and

Marusawa, 2009), where it appears to be not only central to oncogenic transformation (Pasqualucci *et al.*, 2008), but also to tumor relapse after initial therapy (Chiba and Marusawa, 2009; Feldhahn *et al.*, 2007; Mullighan *et al.*, 2008).

1.4 A Unifying Model for AID Function

How can all these observations be reconciled to our current knowledge of AID and its role in the immune system? A clue toward a "unifying model" for AID function is perhaps provided by the hypothesis, first put forth by Petersen-Mahrt (Morgan *et al.*, 2004) that AID can act as a *de facto* active demethylase, i.e. it can deaminate mCpGs yielding TpG, which can be faithfully repaired back to CpG through the action of TDG. A flood of recent experimental evidence supports this notion: AID (with APOBEC–2 and TDG) appears to be at the core of an active demethylation system in zebrafish (Rai *et al.*, 2008); ectopic expression of AID can lead to specific reprogramming toward pluripotency in stem cells (Bhutani *et al.*, 2009); and finally, the primordial germ cell genome in AID-deficient animals appears to be hypermethylated (Popp *et al.*, 2010). Though these recent experiments are by no means flawless, they do provide support for the interesting hypothesis that AID might have evolved to demethylate CpGs in an error-free fashion for purposes of epigenetic reprogramming.

Taken together, the data discussed above suggest that the finding that AID deaminates genes other than Ig genes is not merely due to dysregulation and a by-product of its "real" function. This novel function for AID raises some interesting issues. One is that, though AID can deaminate methyl-dC *in vitro* (Morgan *et al.*, 2004) it has higher activity on dC (Bransteitter *et al.*, 2003; Larijani *et al.*, 2005). Another is that AID activity appears to be directed exclusively toward single-stranded DNA (Bransteitter *et al.*, 2003; Dickerson *et al.*, 2003), whereas methylated DNA would presumably be double-stranded, prior to transcription. However, this difficulty might not be insurmountable as many non-coding gene regions have been shown to be transcribed. In addition, promoter-proximal regions that are dense in CpG motifs, often contain non-canonical DNA structures which expose single-stranded regions (Ball *et al.*, 2009; Tsai *et al.*, 2009; Wittig *et al.*, 1991).

But is there a need for reprogramming toward pluripotency in healthy, functional, adult cells in which AID is expressed? One could argue that germinal center B cells could be just such cells. Germinal centers are hotbeds of epigenetic reprogramming, leading to the differentiation of activated B cells into memory cells and terminal differentiation into long-lived plasma cells. Plasma cells express a very different genetic profile from naïve, germinal center or memory B cells (Bhattacharya *et al.*, 2007; Klein *et al.*, 2003). Perhaps AID-mediated CpG demethylation at non-Ig loci contributes to B cell memory and plasma cell differentiation.

Conversely, there is tremendous pressure toward reprogramming (and eventually pluripotency) in cancers. If we assume that one role for AID is epigenetic reprogramming, and that reprogramming is central to tumorigenesis, then there might be a common signaling pathway that would lead to AID expression under conditions of transformation. One intriguing possibility is that this signal is the switch from mitochondrial oxidative phosphorylation to aerobic glycolysis, also referred to as "the Warburg effect" (Vander Heiden *et al.*, 2009), which allows rapidly dividing cells (such as germinal center B cells as well as all cancer cells) to generate the energy and building blocks needed for growth. Aerobic glycolysis is a universal requirement for cancer cells, and hypoxia (which functionally mimics the Warburg effect) is known to rapidly induce AID expression in cultured B cells (Kim *et al.*, 2006). Of course, each cancerous tissue would have different requirements for reprogramming, which would explain the curious correspondence between the types of AID-promoted translocations seen in different tumors (e.g. *IgH:myc* for B cells [Ramiro *et al.*, 2004]; ETS:TMPRSS2 for prostate cells [Lin *et al.*, 2009]) and the types of stimuli required for these tumors to reprogram toward pluripotency and progress toward malignancy. The expectation then would be that AID targets different genes for demethylation in different settings, and that the selection of translocation partners and their amplification in such settings is lineage-specific.

This year has marked a decade of research into AID and its role in generating antibody diversity. Though we have still a lot to learn in that context, recent work would place AID in a much wider milieu with regard to its contributions to health and disease, and we are looking forward to an integrated understanding of AID function.

1.5 Acknowledgements

We gratefully acknowledge NIH support to the Stavnezer lab (RO1 AI23283) and to the Papavasiliou lab (R01 CA098495).

1.6 References

1. Ball M.P., Li J.B., Gao Y. *et al.* (2009). Targeted and genome-scale strategies reveal gene-body methylation signatures in human cells. *Nat Biotechnol* **27**: 361–368.
2. Besmer E., Market E., Papavasiliou F.N. (2006). The transcription elongation complex directs activation-induced cytidine deaminase-mediated DNA deamination. *Mol Cell Biol* **26**: 4378–4385.
3. Bhattacharya D., Cheah M.T., Franco C.B. *et al.* (2007). Transcriptional profiling of antigen-dependent murine B cell differentiation and memory formation. *J Immunol* **179**: 6808–6819.
4. Bhutani N., Brady J.J., Damian M. *et al.* (2009). Reprogramming towards pluripotency requires AID-dependent DNA demethylation. *Nature* **463**: 1042–1047.
5. Bransteitter R., Pham P., Scharff M.D. *et al.* (2003). Activation-induced cytidine deaminase deaminates deoxycytidine on single-stranded DNA but requires the action of RNase. *Proc Natl Acad Sci USA* **100**: 4102–4107.
6. Chaudhuri J., Tian M., Khuong C. *et al.* (2003). Transcription-targeted DNA deamination by the AID antibody diversification enzyme. *Nature* **422**: 726–730.
7. Chiba T., Marusawa H. (2009). A novel mechanism for inflammation-associated carcinogenesis; an important role of activation-induced cytidine deaminase (AID) in mutation induction. *J Mol Med* **83**: 1023–1027.
8. Conticello S.G., Langlois M.-A., Yang Z. *et al.* (2007). DNA deamination in immunity: AID in the context of its APOBEC relatives. *Adv Immunol* **94**: 37–73.
9. Dickerson S.K., Market E., Besmer E. *et al.* (2003). AID mediates hypermutation by deaminating single stranded DNA. *J Exp Med* **197**: 1291–1296.
10. Feldhahn N., Henke N., Melchior K. *et al.* (2007). Activation-induced cytidine deaminase acts as a mutator in BCR-ABL1-transformed acute lymphoblastic leukemia cells. *J Exp Med* **204**: 1157–1166.

11. Grosjean H., de Crecy-Lagard V. *et al.* (2004). Aminoacylation of the anticodon stem by a tRNA-synthetase paralog: relic of an ancient code? *Trends Biochem Sci* **29**: 519–522.
12. Hamilton C., Papavasiliou F., Rosenberg B. (2010). Diverse functions for DNA and RNA editing in the immune system. *RNA Biol* **7**: 1–10.
13. Imai K., Slupphaug G., Lee W.I. *et al.* (2003). Human uracil-DNA glycosylase deficiency associated with profoundly impaired immunoglobulin class-switch recombination. *Nat Immunol* **4**: 1023–1028.
14. Kim J.W., Tchernyshyov I., Semenza G.L. *et al.* (2006). HIF-1-mediated expression of pyruvate dehydrogenase kinase: a metabolic switch required for cellular adaptation to hypoxia. *Cell Metab* **3**: 177–185.
15. Klein U., Tu Y., Stolovitzky G.A. *et al.* (2003). Transcriptional analysis of the B cell germinal center reaction. *Proc Natl Acad Sci USA* **100**: 2639–2644.
16. Larijani M., Frieder D., Sonbuchner T.M. *et al.* (2005). Methylation protects cytidines from AID-mediated deamination. *Mol Immunol* **42**: 599–604.
17. Lin C., Yang L., Tanasa B. *et al.* (2009). Nuclear receptor-induced chromosomal proximity and DNA breaks underlie specific translocations in cancer. *Cell* **139**: 1069–1083.
18. Liu M., Duke J.L., Richter D.J. *et al.* (2008). Two levels of protection for the B cell genome during somatic hypermutation. *Nature* **451**: 841–845.
19. Martin A., Bardwell P.D., Woo C.J. *et al.* (2002). Activation-induced cytidine deaminase turns on somatic hypermutation in hybridomas. *Nature* **415**: 802–806.
20. Mayorov V.I., Rogozin I.B., Adkison L.R. *et al.* (2005). Expression of human AID in yeast induces mutations in context similar to the context of somatic hypermutation at G-C pairs in immunoglobulin genes. *BMC Immunol* **6**: 10.
21. Morgan H.D., Dean W., Coker H.A. *et al.* (2004). Activation-induced cytidine deaminase deaminates 5-methylcytosine in DNA and is expressed in pluripotent tissues: implications for epigenetic reprogramming. *J Biol Chem* **279**: 52353–52360.
22. Mullighan C.G., Phillips L.A., Su X. *et al.* (2008). Genomic analysis of the clonal origins of relapsed acute lymphoblastic leukemia. *Science* **322**: 1377–1380.
23. Muramatsu M., Kinoshita K., Fagarasan S. *et al.* (2000). Class switch recombination and hypermutation require activation-induced cytidine deaminase (AID), a potential RNA editing enzyme. *Cell* **102**: 553–563.
24. Muramatsu M., Sankaranand V.S., Anant S. *et al.* (1999). Specific expression of activation-induced cytidine deaminase (AID), a novel member of the RNA-editing deaminase family in germinal center B cells. *J Biol Chem* **274**: 18470–18476.
25. Nakamura M., Kondo S., Sugai M. *et al.* (1996). High frequency class switching of an IgM+ B lymphoma clone CH12F3 to IgA+ cells. *Int Immunol* **8**: 193–201.
26. Pasqualucci L., Bhagat G., Jankovic M. *et al.* (2008). AID is required for germinal center-derived lymphomagenesis. *Nat Genet* **40**: 108–112.
27. Pasqualucci L., Neumeister P., Goossens T. *et al.* (2001). Hypermutation of multiple proto-oncogenes in B-cell diffuse large-cell lymphomas. *Nature* **412**: 341–346.

28. Petersen-Mahrt S.K., Harris R.S., Neuberger M.S. (2002). AID mutates E. coli suggesting a DNA deamination mechanism for antibody diversification. *Nature* **418**: 99–103.
29. Pham P., Bransteitter R., Petruska J. *et al.* (2003). Processive AID-catalysed cytosine deamination on single-stranded DNA simulates somatic hypermutation. *Nature* **424**: 103–107.
30. Poltoratsky V.P., Wilson S.H., Kunkel T.A. *et al.* (2004). Recombinogenic phenotype of human activation-induced cytosine deaminase. *J Immunol* **172**: 4308–4313.
31. Popp C., Dean W., Feng S. *et al.* (2010). Genome-wide erasure of DNA methylation in mouse primordial germ cells is affected by AID deficiency. *Nature* **463**: 1101–1105.
32. Rada C., Di Noia J.M., Neuberger, M.S. (2004). Mismatch recognition and uracil excision provide complementary paths to both Ig switching and the A/T-focused phase of somatic mutation. *Mol Cell* **16**: 163–171.
33. Rada C., Williams G.T., Nilsen H. *et al.* (2002). Immunoglobulin isotype switching is inhibited and somatic hypermutation perturbed in UNG-deficient mice. *Curr Biol* **12**: 1748–1755.
34. Rai K., Huggins I.J., James S.R. *et al.* (2008). DNA demethylation in zebrafish involves the coupling of a deaminase, a glycosylase, and gadd45. *Cell* **135**: 1201–1212.
35. Ramiro A.R., Jankovic M., Eisenreich T. *et al.* (2004). AID is required for c-myc/IgH chromosome translocations in vivo. *Cell* **118**: 431–438.
36. Revy P., Muto T., Levy Y. *et al.* (2000). Activation-induced cytidine deaminase (AID) deficiency causes the autosomal recessive form of the Hyper-IgM syndrome (HIGM2). *Cell* **102**: 565–575.
37. Shen H.M., Peters A., Baron B. *et al.* (1998). Mutation of BCL-6 gene in normal B cells by the process of somatic hypermutation of Ig genes. *Science* **280**: 1750–1752.
38. Shen H.M., Poirier M.G., Allen M.J. *et al.* (2009). The activation-induced cytidine deaminase (AID) efficiently targets DNA in nucleosomes but only during transcription. *J Exp Med* **206**: 1057–1071.
39. Tsai A.G., Engelhart A.E., Hatmal M.M. *et al.* (2009). Conformational variants of duplex DNA correlated with cytosine-rich chromosomal fragile sites. *J Biol Chem* **284**: 7157–7164.
40. Vander Heiden M.G., Cantley L.C., Thompson C.B. (2009). Understanding the Warburg effect: the metabolic requirements of cell proliferation. *Science* **324**: 1029–1033.
41. Wang M., Rada C., Neuberger M.S. (2010) Altering the spectrum of immunoglobulin V gene somatic hypermutation by modifying the active site of AID. *J Exp Med* **207**: 141–153.
42. Wedekind J.E., Dance G.S.C., Sowden M.P. *et al.* (2003). Messenger RNA editing in mammals: new members of the APOBEC family seeking roles in the family business. *Trends Genet* **19**: 207–216.

43. Wittig B., Dorbic T., Rich A. (1991). Transcription is associated with Z-DNA formation in metabolically active permeabilized mammalian cell nuclei. *Proc Natl Acad Sci USA* **88**: 2259–2263.
44. Yoshikawa K., Okazaki I.M., Eto T. *et al.* (2002). AID enzyme-induced hypermutation in an actively transcribed gene in fibroblasts. *Science* **296**: 2033–2036.

Chapter 2

Switch Regions, Chromatin Accessibility and AID Targeting

Amy L. Kenter[1] and Patricia J. Gearhart[2]

[1]Department of Microbiology and Immunology
University of Illinois College of Medicine
Chicago, IL 60612-7344, USA
E-mail: starl@uic.edu

[2]Laboratory of Molecular Gerontology
National Institute on Aging / National Institutes of Health
Baltimore, MD21224, USA 21224
E-mail: gearhartp@mail.nih.gov

During B cell activation, activation-induced deaminase (AID) is targeted to switch (S) regions and variable regions but not to other loci. The mechanism regulating the differential targeting of AID to its cognate substrates remains unclear. Although it has long been known that transcription is critically required for class switch recombination and somatic hypermutation, it has not been clear whether or by what means transcription facilitates AID targeting. Here, we consider how transcription across S DNA leads to high occupancy of RNA polymerase II, which in turn promotes the introduction of activating chromatin modifications and highly accessible chromatin that is open to AID attack. Of considerable interest is data indicating that the unique structure of the S regions is mechanistically linked to generating open chromatin. These advances have led to the recognition of a nexus between transcription and long-range intra-chromosomal interactions among IgH transcriptional elements, chromatin remodeling and histone modifications.

2.1 Introduction

In mature B cells, class switch recombination (CSR) promotes diversification of IgH effector functions encoded in constant (C_H) regions while maintaining the original antigen binding specificity arising from V(D)J recombination. A mutational process termed somatic hypermutation (SHM), is focused to rearranged V_HDJ_H region genes in germinal center B cells and leads to affinity maturation of antibody binding to antigen. The mouse immunoglobulin heavy chain (*Igh*) locus contains eight constant (C_H) genes (μ, δ, γ3, γ1, γ2a, γ2b, ε and α) that are located downstream of the V, D and J_H segments and each C_H region is paired with a complementary switch (S) region (with the exception of Cδ). CSR occurs between S region sequences leading to an intra-chromosomal deletional rearrangement that results in the formation of composite Sμ–Sx junctions on the chromosome, while the intervening genomic material is looped out and excised. Activation-induced deaminase (AID) is the master regulator of CSR and SHM (Muramatsu *et al.*, 2000). SHM requires AID for production of mutations targeted to the expressed V genes at rates that are orders of magnitude higher than background. During CSR, AID initiates formation of S region-specific double-strand breaks (DSBs) that are processed by a cascade of events mediated by nonhomologous end joining (NHEJ). The mechanism of AID deamination leading to DNA lesions that initiate SHM and CSR with subsequent DNA repair, have been extensively reviewed and will not be expanded upon here (Peled *et al.*, 2008; Martin and Scharff, 2002a; Di Noia and Neuberger, 2002; Neuberger *et al.*, 2005; Martomo and Gearhart, 2006; Saribasak *et al.*, 2009; Chaudhuri and Alt, 2004; Chaudhuri *et al.*, 2007; Teng and Papavasiliou, 2007; Stavnezer *et al.*, 2008; Dudley *et al.*, 2005).

One of the most puzzling aspects of AID behavior is the preferential focus on Ig genes despite its capacity to deaminate any *in vitro* transcribed ssDNA substrate (Dickerson *et al.*, 2003; Chaudhuri *et al.*, 2003; Pham *et al.*, 2003; Shen and Storb, 2004; Ramiro *et al.*, 2003). AID also appears to target a select set of highly transcribed genes when it is highly expressed in germinal center or Peyers patch B cells (Liu *et al.*, 2008) or upon ectopic overexpression (Martin *et al.*, 2002; Martin and

Scharff, 2002b; Okazaki *et al.*, 2003; Yoshikawa *et al.*, 2002). Similarly it has been observed that loss of AID target specificity might be a consequence of environmental events that boost AID expression (Rosenberg and Papavasiliou, 2007). Alternatively, it has been suggested that AID mistargeting occurs frequently in highly transcribed genes in normal B cells but the emergence of mutations occurs through a breakdown of high fidelity repair (Liu *et al.*, 2008), a perturbation of normal error-prone DNA polymerase expression (Epeldegui *et al.*, 2007; Machida *et al.*, 2005) or mismatch repair proteins (Bindra and Glazer, 2007) that are implicated in repair of AID-induced lesions. Poorly controlled AID targeting has been postulated to underlie aberrant SHM of non-Ig genes in B cell tumors (Pasqualucci *et al.*, 2001), *Igh:c-myc* translocations in mature B cells undergoing CSR (Kuppers and Dalla-Favera, 2001; Takizawa *et al.*, 2008) and induction of AID in human gastric epithelial cells following infection with *Heliobacter pylori* leading to mutations in TP53 (Matsumoto *et al.*, 2007). In this regard, one of the most pressing challenges in the field is determining the mechanism by which AID is targeted to Ig loci.

A hallmark feature of CSR and SHM is the requirement for transcription and transcriptional elements as indicated by gene-targeting experiments (Delker *et al.*, 2009; Stavnezer, 2000; Perlot and Alt, 2008). Eukaryotic promoters are located immediately upstream of transcription start sites and provide a platform for RNA polymerase recruitment. As shown in Fig. 2.1, the rearranged V_HDJ_H exon is located upstream of the Cμ exons and a promoter 5′ of the V_H segment drives transcription. SHM is initiated by transcription-dependent targeting of AID to the

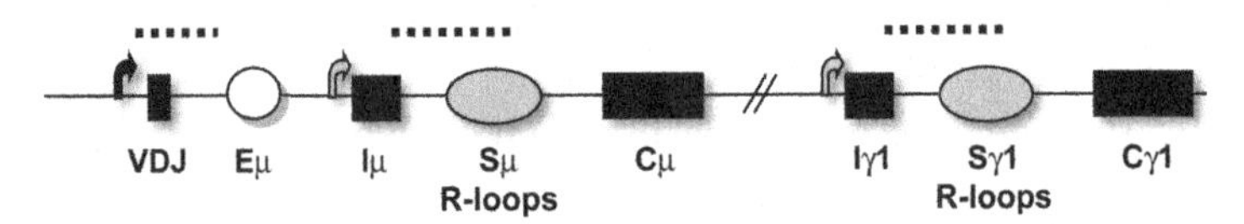

Figure 2.1. **Focus of AID activity in the *Igh* locus.** The diagram shows exons (black boxes) encoding VDJ, Iμ, and Cμ genes, and some distance away, the Iγ1 and Cγ1 genes. The intronic μ enhancer, Eμ, is indicated by an open circle, and Sμ and Sγ1 regions by gray ovals. Bent arrows represent the VDJ promoter (black) and I promoters (gray). Dotted lines above the drawing show the extent of AID activity as measured by detection of mutations.

mature V_HDJ_H exon, followed by error-prone repair of the resulting AID-induced mismatches (Di Noia and Neuberger, 2007). These and other observations suggest that transcription *per se* or the transcription apparatus is a prerequisite for the process of SHM (Goyenechea *et al.*, 1997; Tumas-Brundage and Manser, 1997; Fukita *et al.*, 1998). Regarding CSR, each S region is paired with an I (intervening) exon and its associated promoter. Transcription initiates at the 5′ end of the I exon, proceeds through the S region and terminates downstream of the corresponding C_H gene (Chaudhuri *et al.*, 2007). Transcription from different I exon promoters is induced by specific combinations of cytokines and B cell activators and targets a S region for CSR (Stavnezer, 2000). The transcribed mRNA products are termed sterile or germline transcripts (GLTs) because they do not contain an open reading frame and there is no evidence that they are translated (Chaudhuri *et al.*, 2007). Gene-targeting studies have shown a critical requirement for GLTs as a prerequisite for CSR (Manis *et al.*, 2002). The requirement for transcription is a common nexus for SHM and CSR and a crucial prerequisite for AID attack. In the following sections we explore recent insights regarding the mechanism of transcription, its influence on three-dimensional chromatin organization, its relationship to epigenetic chromatin modifications and how these processes integrate to regulate CSR and AID behavior.

2.2 Transcriptional Elements Determine Long-Range Regulation of CSR

Regulatory elements that modulate gene expression include enhancers, silencers, locus control regions (LCR), insulator/boundary elements and matrix attachment sites. *Cis*-regulatory elements influence promoter-directed transcription and are located up to hundreds of kilobases from the genes they regulate. Indeed, S regions targeted for recombination can be separated by as much as 150kb, and this distance is likely to be an impediment for S-S synapse formation. This observation leads to the perplexing question of how distantly located S-region-specific DSBs are recruited to partner in a CSR reaction that leads to intra-chromosomal deletion. The development of chromosome conformation capture (3C) techniques permits measurement of physical interactions between chromatin fibers (Dekker *et al.*, 2002; Miele and Dekker, 2009). The mouse β-globin genes and their LCR located more than 50kb away interact to form higher order loop structures, and these interactions are tightly correlated with gene-specific transcription (Carter *et al.*, 2002; Tolhuis *et al.*, 2002). Loop structures have been detected in numerous complex mutagenic loci including imprinted loci (Horike *et al.*, 2005; Murrell *et al.*, 2004), MHC class II (Majumder *et al.*, 2008), cytokine clusters (Spilianakis and Flavell, 2004; Sekimata *et al.*, 2009) and between chromatin boundary elements (Blanton *et al.*, 2003; Phillips and Corces, 2009).

In the *Igh* locus, inducible transcription from the downstream GLT promoters requires the 3′Eα LCR (Manis *et al.*, 1998; Pinaud *et al.*, 2001), suggesting that long-range intra-chromosomal interactions facilitate communication between these distant *cis* elements and lead to loop formation and transcription. However, the spatial proximity between the downstream GLT promoters with 3′Eα would not bring the downstream S regions into proximity with Sμ and could not bring about S-S synapsis. Studies using 3C assays demonstrate that the *Igh* locus assumes a unique chromosomal loop conformation *in vivo* in which the area around the Eμ enhancer directly interacts with the downstream 3′Eα LCR (Wuerffel *et al.*, 2007; Ju *et al.*, 2007; Sellars *et al.*, 2009). Since the Sμ region is closely arrayed in *cis* with Eμ, the Eμ:3′Eα interaction could facilitate S-S synapsis following GLT promoter activation as a downstream S region travels with its proximal GLT promoter to the Eμ:3′Eα complex. Strong recruitment of GLT promoters to the Eμ:3′Eα complex was cytokine-dependent and correlated with transcription activation (Wuerffel *et al.*, 2007). These interactions are dependent on the 3′Eα LCR as deletion of the hypersensitive site (hs) 3b,4 largely abolishes Eμ:3′Eα interactions with each other and the GLT promoter, whereas deletion of the Sμ tandem repeats had little impact (Wuerffel *et al.*, 2007). Deletion of the 220bp core Eμ enhancer had a modest effect on recruitment of GLT promoters to the Eμ:3′Eα and only a slight effect on CSR, indicating that the critical *cis* element required for these interactions has not yet been identified (Wuerffel *et al.*, 2007). Detection of *Igh* locus-wide looping has led to a new model for generating S-S synapsis in which GLT expression is integrally linked to the formation of a three-dimensional architectural scaffold produced through long-range associations between *Igh* transcription regulatory elements.

2.3 *Cis*-Regulatory Elements as Recruiters for AID

The search for *cis*-regulatory elements involved in targeting AID to the *Igh* locus as distinct from mediating transcription has been difficult, with two exceptions. Earlier studies indicated that promoters are fungible for both SHM and CSR where heterologous promoters functioned well in both formats (Jung *et al.*, 1993; Martin and Scharff, 2002b; Lorenz *et al.*, 1995). Recent evidence suggests that not all promoters are equal in their capacity to support SHM. In the DT40 chicken B cell line, substitution of the endogenous Igλ with the highly transcribed elongation factor 1-α promoter leads to a reduction in mutagenesis, suggesting that promoter identity impacts on SHM efficiency (Yang *et al.*, 2006). Using deletional analysis in DT40 cells, a mutational enhancer was identified that confers mutability on heterologous genes and is distinct from the known transcriptional enhancer elements (Blagodatski *et al.*, 2009; Kothapalli *et al.*, 2008). These studies are the first to delineate an AID recruitment element, although the precise DNA motifs responsible for the effect remain obscure and similar findings remain elusive in mouse and human.

2.4 Transcription and Accessibility to AID Attack

Eukaryotic DNA is wrapped around histone octamers to form nucleosomes which in turn are organized into higher-order chromatin fibers to preclude access of *trans*-acting factors to DNA (Woodcock and Dimitrov, 2001). The specificity of AID for single-stranded (ss) DNA templates (Chaudhuri *et al.*, 2007) requires that S region substrates become accessible in chromatin prior to CSR. As the transcription complex moves through chromatin, nucleosomes are displaced and DNA becomes accessible (Belotserkovskaya *et al.*, 2004). However, it is puzzling that while S and C_H regions are located within the same transcriptional unit, S regions are the preferred target for AID deamination.

A new paradigm for transcription regulation has emerged that incorporates the phosphorylation status of RNA polymerase II (RNAP II) C-terminal domain (CTD) with recruitment of histone-modifying enzymes, which in turn introduce histone marks that alter the status of chromatin accessibility (Saunders *et al.*, 2006; Li *et al.*, 2007). Genome-wide analyses indicate that in transcriptionally-active genes, promoter proximal sites are enriched with histones that are hyper-acetylated (Ac), trimethylated on histone H3 at lysine 4 (H3K4me3), and with RNAP II phosphorylated at serine 5 (p-ser5), whereas levels of H3K36me3 and elongating RNAP II p-ser2 are elevated in the downstream coding regions of active genes (Bernstein *et al.*, 2005; Pokholok *et al.*, 2005; Barski *et al.*, 2007; Guenther *et al.*, 2007).

What are the consequences of localizing H3K4me3 to promoter proximal regions and H3K36me3 to the downstream coding regions? Studies indicate that H3K4me3 is directly bound by a constituent of the NuA3 histone acetytransferase (HAT) complex that coordinates transcription activation with histone Ac (Taverna *et al.*, 2007; Berger, 2007). Histone Ac may change the folding properties of the chromatin fiber and alter the net charge of nucleosomes, resulting in increased DNA accessibility (Shahbazian and Grunstein, 2007). Histone Ac may also create binding surfaces for specific protein-histone interactions which then provide a versatile code for recruitment of factors to control transcription (Li *et al.*, 2007). H3K36me3, in yeast, is recognized by the Rpd3S histone deactylase (HDAC) complex, which functions to reduce histone Ac and suppress inappropriate transcription initiation in downstream coding regions (Carrozza *et al.*, 2005; Lieb and Clarke, 2005). Thus, the underlying H3K4me3 and H3K36me3 methylation pattern provides for recruitment of HATs and HDACs, respectively, which in turn reciprocally regulate chromatin accessibility to assure spatially appropriate transcription initiation within the transcription unit.

Paradoxically, targeting of S regions by AID is transcription-dependent whereas C_H regions within the same transcription unit are protected from AID attack. In this regard, it is significant that upon transcription activation, I-S regions are a focus for increased H3K4me3 while C_H regions accumulate the repressive countermarks, H3K36me3 and H4K20me1 (Wang *et al.*, 2009). S regions are a focus for histone Ac (Wang *et al.*, 2006; Nambu *et al.*, 2003; Li *et al.*, 2004) and chromatin hyperaccessibility (Wang *et al.*, 2009), whereas C_H regions remain relatively hypoAc and are inaccessible in chromatin (Wang *et al.*, 2009; Wang *et al.*, 2006). Indeed, antisense RNA transcripts are detected in S regions but not in C_H regions, reflecting the reciprocal zones of accessible or inaccessible chromatin (Perlot and Alt, 2008). Modulation of histone Ac levels using trichostatin A leads to increased histone Ac and chromatin accessibility as well as elevated CSR frequency, indicating the physiological importance of these chromatin modifications (Wang *et al.*, 2009). Histone Ac status and chromatin accessibility are most likely to be patterned across I-S-C_H loci by means of the underlying distribution of H3K4me3 and H3K36me3 modifications, which are in turn integrally linked to transcription. This model draws from the established observations that H3K4me3 is a substrate for HAT binding whereas H3K36me3 recruits HDAC (Carrozza *et al.*, 2005; Lieb and Clarke, 2005; Taverna *et al.*, 2007; Berger, 2007). Thus, reciprocally accessible and repressed chromatin environments coincide with S regions and C_H regions, respectively, and are inherently linked to the mechanism of transcription.

H3K9me3 is another histone mark detected in transcribed S regions but absent from C_H regions (Kuang *et al.*, 2009). This is a curious observation since H3K9me3 is considered a repressive modification most frequently associated with silent pericentromeric heterochromatin (Berger, 2007). The combination of H3K4me3 (activating) and H3K9me3 (repressive) marks have been detected across the body of actively-transcribed genes (Vakoc *et al.*, 2005; Vakoc *et al.*, 2006), providing precedence for detection of H3K9me3 in actively-transcribed S regions. H3K9me3 modifications are introduced by several histone methyltransferases (HMTs) including Suv39h1 (Kouzarides, 2007). Overexpression of Suv39h1 leads to increased $\mu \rightarrow \alpha$ CSR on a switch plasmid but does not function to enhance CSR for other switch substrate isotypes (Bradley *et al.*, 2006). Mutation of the Suv39h1 SET domain abrogates HMT activity and abolishes $\mu \rightarrow \alpha$ switch plasmid CSR

(Bradley *et al.*, 2006). Endogenous switching μ–>α is selectively reduced in Suv39h1 deficient B cells, indicating an isotype-specific influence of Suv39h1 on CSR and corroborates the switch plasmid findings (Bradley *et al.*, 2006). Detection of H3K9me3 in transcribed S regions suggests that the influence of Suv39h1 is direct and specific for IgA switching. The underlying mechanism by which H3K9me3 and its HMTs influence CSR requires further investigation.

2.5 S Region Sequence Determines Chromatin Accessibility

S regions extend 1 to 10kb downstream of the I exons (Gritzmacher, 1989), and when transcribed are decorated with activating histone modifications H3Ac, H4Ac and H3K4me3. They are also hyperaccessible throughout their lengths (Wang *et al.*, 2006; Wang *et al.*, 2009). However, genome-wide analyses in yeast and humans show that these modifications are generally restricted to promoter proximal locations and rarely extend more that 1kb downstream of the promoter (Bernstein *et al.*, 2005; Pokholok *et al.*, 2005). What directs the spread of these modifications throughout S regions and long distances from the GLT promoters? The H3K4 HMT binds to initiating RNAP II p-ser 5 and introduces H3K4me3 into promoter proximal regions (Hampsey and Reinberg, 2003). To determine whether a similar mechanism functions in S DNA, Wang and colleagues used ChIP analyses to find high-density RNAP II p-ser 5 spread throughout two actively-transcribed S regions (Wang *et al.*, 2009). Independently, Rajagopal and colleagues also detected high occupancy RNAP II in the Sμ region, using ChIP and nuclear run-on assays (Rajagopal *et al.*, 2009). In B cells devoid of Sμ DNA, RNAP II occupancy (Rajagopal *et al.*, 2009; Wang *et al.*, 2009) and H3K4me3 (Wang *et al.*, 2009) remained promoter proximal, indicating that S region sequence facilitates retention of RNAP II p-ser5 and is responsible for the extended zone of H3K4me3 modification, highlighting unusual properties for S region sequence.

Mammalian S regions are uniquely rich with clusters of G nucleotides on the nontranscribed strand, and repetitive hotspot motifs for AID deamination, WGCW (W = A or T; Stavnezer and Amemiya, 2004). In transcribed S regions, the G residue clusters are responsible for R-loop formation which is generated by hybridization of the nascent RNA transcript with the transcribed DNA template, while the nontranscribed DNA strand is exposed as ssDNA (Yu *et al.*, 2003; Huang *et al.*, 2007). R loops in S regions (Fig. 2.1) are thought to provide ssDNA stretches as substrate for AID deamination and thereby enhance the efficiency of CSR (Yu *et al.*, 2003; Huang *et al.*, 2007). S regions may also be subject to formation of G-loops which form from the combination of G-quartet and R-loop structures (Tornaletti *et al.*, 2008; Duquette *et al.*, 2004). However, G-loops have been not been observed in S regions undergoing CSR *in vivo* (Duquette *et al.*, 2004). Strikingly, RNA-DNA hybrid structures also impede transcription elongation (Tous and Aguilera, 2007; Huertas and Aguilera, 2003) and may be responsible for the high-density RNAP II observed in transcriptionally-active S regions (Rajagopal *et al.*, 2009; Wang *et al.*, 2009). However, a causal relationship between R-loops and RNAP II enrichment awaits direct experimental demonstration.

2.6 AID-Induced Mutation Distribution and Transcription

The mutation distribution across the V exons has a sharp 5′ boundary positioned approximately 120bp downstream of the transcription start sites and a less demarcated 3′ boundary about 1kb beyond the promoter (Lebecque and Gearhart, 1990; Rada *et al.*, 1997). Alteration of the promoter position leads to a corresponding displacement of transcription initiation and perturbation of mutation distribution (Peters and Storb, 1996; Bachl *et al.*, 2001), strongly implicating promoter proximal transcription with inducing AID-dependent mutations. Based on these observations Peters and Storb postulated that a mutator (now known to be AID) associates with the transcription initiation apparatus, tracks with the transcription elongation complex and dissociates stochastically to produce mutations (Peters and Storb, 1996). This model was later extended to CSR because mutation distribution begins 150bp downstream of the I exon transcription start site, similar to the pattern found for V exons (Xue *et al.*, 2006). The 3′ boundary for CSR mutations is located downstream of S regions, a distance of up to 10kb from the transcription start site in agreement with the patterns of chromatin accessibility established for the S region and track with high-density RNAP II (Gritzmacher, 1989; Xue *et al.*, 2006; Rajagopal *et al.*, 2009; Wang *et al.*, 2009). This hypothesis has gained experimental support from the findings that ectopically-expressed AID binds with RNAP II in co-immunoprecipitation assays (Nambu *et al.*, 2003) and interacts with the transcription apparatus *in vitro* (Besmer *et al.*, 2006). However, additional studies using more physiologically relevant systems are required for confirmation.

2.7 Processing of GLTs and the Introduction of AID-Induced Mutations

AID-induced deamination of dC residues leading to DSB formation in the S regions is dependent on GLT expression (Chaudhuri *et al.*, 2007; Stavnezer *et al.*, 2008), and GL transcription functions exclusively in *cis* during CSR (Bottaro *et al.*, 1994). The interpretation of similar analyses focused on the μ GLT are complicated by the presence of alternative Iμ exons which function redundantly (Kuzin *et al.*, 2000). The noncoding GLT is spliced to join the I exon to the downstream C_H exon, thereby removing the S region sequences (Chaudhuri *et al.*, 2007). Unexpectedly, deletion of the γ1 I exon splice donor abolished μ–>γ1 CSR (Hein *et al.*, 1998; Lorenz *et al.*, 1995). Recently, CTNNBL1, a spliceosome associated factor, was shown to interact with AID *in vivo* and *in vitro*, and disruption of this interaction in DT40 cells reduces IgV diversification (Conticello *et al.*, 2008). A point mutation in AID abolishes interaction with CTNNBL1 and leads to reduced CSR in murine B cells (Conticello *et al.*, 2008). It is noteworthy that transcription, mRNA processing and splicing occur contemporaneously (McCracken *et al.*, 1997; Johnson *et al.*, 2009). Additionally, mutation frequencies by AID are highly elevated in yeast depleted for the THO-TREX complex (Gomez-Gonzalez and Aguilera, 2007). The THO-TREX complex is centrally involved in transcription and mRNA splicing in yeast (Gomez-Gonzalez and Aguilera, 2007) and mammalian cells

(Masuda *et al.*, 2005). The biologically significant interaction between AID and CTNNBL1 potentially provides a mechanistic link between the requirement for the splicing of GLTs and recruitment of AID to the transcription apparatus, thereby potentiating CSR, although it remains unclear whether CTNNBL1 provides specificity for AID engagement with Ig genes. The association of AID with CTNNBL1 at the I exon splice donor would position AID to initiate deamination events precisely where mutations are first detected in the transcription unit.

2.8 Future Directions

The mechanism by which AID is focused primarily to Ig loci (V and S regions) during SHM and CSR to the exclusion of other transcriptionally-active genes remains the most intriguing problem in the field. It is currently clear that poorly understood processes organize higher-order chromatin structures in three-dimensional nuclear space to regulate transcriptional programs and patterns of replication (Cook, 1999). For example, large heterochromatic regions assemble near the nuclear lamina whereas euchromatin loops locate to the interior of the nucleus suggesting that the gene positioning determines gene expression status (Zhao *et al.*, 2009). Distantly located genes positioned on different chromosomes can converge in transcription factories (Osborne *et al.*, 2004: Osborne *et al.*, 2007) or at nuclear speckles which contain high concentrations of splicing and transcription elongation factors (Pandit *et al.*, 2008; Lamond and Spector, 2003). Nuclear pore complexes are interspersed in the nuclear membrane to allow nucleo-cytoplasmic exchange (Tran and Wente, 2006), are permissive for transcription and coordinate RNA processing in yeast (Taddei *et al.*, 2006; Rougemaille *et al.*, 2008). One solution to the enigma of AID targeting could be that AID and Ig genes are coordinately sequestered to the same transcriptionally-permissive nuclear subcompartment to the exclusion of non-Ig genes. A particularly intriguing hypothesis in this regard has AID and Ig genes in identical nuclear pore complexes since AID is both imported to, and exported from, the nucleus and its export signal is critical for function (Delker *et al.*, 2009). Another, equally plausible idea is that when Ig genes are nonrandomly distributed in transcription factories (Osborne *et al.*, 2007), the splicing factor CTTNBL1 is concentrated there with its interaction partner AID to produce focused mutagenesis. The next iteration of AID studies is likely to be informative and exciting in equal measure.

2.9 Acknowledgements

This work was supported in part by the National Institutes of Health (AI052400 to A.L.K.) and the Intramural Research Program of the NIH, National Institute on Aging (P.J.G.). The authors have no competing financial interests.

2.10 References

1. Bachl J., Carlson C., Gray-Schopfer V. *et al.* (2001). Increased transcription levels induce higher mutation rates in a hypermutating cell line. *J Immunol* **166**: 5051–5057.
2. Barski A., Cuddapah S., Cui K. *et al.* (2007). High-resolution profiling of histone methylations in the human genome. *Cell* **129**: 823–837.
3. Belotserkovskaya R., Saunders A., Lis J.T. *et al.* (2004). Transcription through chromatin: understanding a complex FACT. *Biochim Biophys Acta* **1677**: 87–99.
4. Berger S.L. (2007). The complex language of chromatin regulation during transcription. *Nature* **447**: 407–412.
5. Bernstein B.E., Kamal M., Lindblad-Toh K. *et al.* (2005). Genomic maps and comparative analysis of histone modifications in human and mouse. *Cell* **120**: 169–181.
6. Besmer E., Market E., Papavasiliou F.N. (2006). The transcription elongation complex directs activation-induced cytidine deaminase-mediated DNA deamination. *Mol Cell Biol* **26**: 4378–4385.
7. Bindra R.S., Glazer P.M. (2007). Co-repression of mismatch repair gene expression by hypoxia in cancer cells: role of the Myc/Max network. *Cancer Lett* **252**: 93–103.
8. Blagodatski A., Batrak V., Schmidl S. *et al.* (2009). A cis-acting diversification activator both necessary and sufficient for AID-mediated hypermutation. *PLoS Genet* **5**: e1000332.
9. Blanton J., Gaszner M., Schedl P. (2003). Protein:protein interactions and the pairing of boundary elements in vivo. *Genes Dev* **17**: 664–675.

10. Bottaro A., Lansford R., Xu L. *et al.* (1994). S region transcription per se promotes basal IgE class switch recombination but additional factors regulate the efficiency of the process. *EMBO J* **13**: 665–674.
11. Bradley S.P., Kaminski D.A., Peters A.H. *et al.* (2006). The histone methyltransferase Suv39h1 increases class switch recombination specifically to IgA. *J Immunol* **177**: 1179–1188.
12. Carrozza M.J., Li B., Florens L. *et al.* (2005). Histone H3 methylation by Set2 directs deacetylation of coding regions by Rpd3S to suppress spurious intragenic transcription. *Cell* **123**: 581–592.
13. Carter D., Chakalova L., Osborne C.S. *et al.* (2002). Long-range chromatin regulatory interactions in vivo. *Nat Genet* **32**: 623–626.
14. Chaudhuri J., Alt F.W. (2004). Class-switch recombination: interplay of transcription, DNA deamination and DNA repair. *Nat Rev Immunol* **4**: 541–552.
15. Chaudhuri J., Basu U., Zarrin A. *et al.* (2007). Evolution of the immunoglobulin heavy chain class switch recombination mechanism. *Adv Immunol* **94**: 157–214.
16. Chaudhuri J., Tian M., Khuong C. *et al.* (2003). Transcription-targeted DNA deamination by the AID antibody diversification enzyme. *Nature* **422**: 726–730.
17. Conticello S.G., Ganesh K., Xue K. *et al.* (2008). Interaction between antibody-diversification enzyme AID and spliceosome-associated factor CTNNBL1. *Mol Cell* **31**: 474–484.
18. Cook P.R. (1999). The organization of replication and transcription. *Science* **284**: 1790–1795.
19. Dekker J., Rippe K., Dekker M. *et al.* (2002). Capturing chromosome conformation. *Science* **295**: 1306–1311.
20. Delker R.K., Fugmann S.D., Papavasiliou F.N. (2009). A coming-of-age story: activation-induced cytidine deaminase turns 10. *Nat Immunol* **10**: 1147–1153.
21. Di Noia J.M., Neuberger M.S. (2002). Altering the pathway of immunoglobulin hypermutation by inhibiting uracil-DNA glycosylase. *Nature* **419**: 43–48.
22. Di Noia J.M., Neuberger M.S. (2007). Molecular mechanisms of antibody somatic hypermutation. *Ann Rev Biochem* **76**: 1–22.
23. Dickerson S.K., Market E., Besmer E. *et al.* (2003). AID mediates hypermutation by deaminating single stranded DNA. *J Exp Med* **197**: 1291–1296.
24. Dudley D.D., Chaudhuri J., Bassing C.H. *et al.* (2005). Mechanism and control of V(D)J recombination versus class switch recombination: similarities and differences. *Adv Immunol* **86**: 43–112.
25. Duquette M.L., Handa P., Vincent J.A. *et al.* (2004). Intracellular transcription of G-rich DNAs induces formation of G-loops, novel structures containing G4 DNA. *Genes Dev* **18**: 1618–1629.
26. Epeldegui M., Hung Y.P., Mcquay A. *et al.* (2007). Infection of human B cells with Epstein-Barr virus results in the expression of somatic hypermutation-inducing molecules and in the accrual of oncogene mutations. *Mol Immunol* **44**: 934–942.
27. Fukita Y., Jacobs H., Rajewsky K. (1998). Somatic hypermutation in the heavy chain locus correlates with transcription. *Immunity* **9**: 105–114.

28. Gomez-Gonzalez B., Aguilera A. (2007). Activation-induced cytidine deaminase action is strongly stimulated by mutations of the THO complex. *Proc Natl Acad Sci USA* **104**: 8409–8414.
29. Goyenechea B., Klix N., Yelamos J. *et al.* (1997). Cells strongly expressing Ig(kappa) transgenes show clonal recruitment of hypermutation: a role for both MAR and the enhancers. *EMBO J* **16**: 3987–3994.
30. Gritzmacher C.A. (1989). Molecular aspects of heavy-chain class switching. *Crit Rev Immunol* **9**: 173–200.
31. Guenther M.G., Levine S.S., Boyer L.A. *et al.* (2007). A chromatin landmark and transcription initiation at most promoters in human cells. *Cell* **130**: 77–88.
32. Hampsey M., Reinberg D. (2003). Tails of intrigue: phosphorylation of RNA polymerase II mediates histone methylation. *Cell* **113**: 429–432.
33. Hein K., Lorenz M.G., Siebenkotten G. *et al.* (1998). Processing of switch transcripts is required for targeting of antibody class switch recombination. *J Exp Med* **188**: 2369–2374.
34. Horike S., Cai S., Miyano M. *et al.* (2005). Loss of silent-chromatin looping and impaired imprinting of DLX5 in Rett syndrome. *Nat Genet* **37**: 31–40.
35. Huang F.T., Yu K., Balter B.B. *et al.* (2007). Sequence dependence of chromosomal R-loops at the immunoglobulin heavy-chain Smu class switch region. *Mol Cell Biol* **27**: 5921–5932.
36. Huertas P., Aguilera A. (2003). Cotranscriptionally formed DNA:RNA hybrids mediate transcription elongation impairment and transcription-associated recombination. *Mol Cell* **12**: 711–721.
37. Johnson S.A., Cubberley G., Bentley D.L. (2009). Cotranscriptional recruitment of the mRNA export factor Yra1 by direct interaction with the 3' end processing factor Pcf11. *Mol Cell* **33**: 215–226.
38. Ju Z., Volpi S.A., Hassan R. *et al.* (2007). Evidence for physical interaction between the immunoglobulin heavy chain variable region and the 3' regulatory region. *J Biol Chem* **282**: 35169–35178.
39. Jung S., Rajewsky K., Radbruch A. (1993). Shutdown of class switch recombination by deletion of a switch region control element. *Science* **259**: 984–987.
40. Kothapalli N., Norton D.D., Fugmann S.D. (2008). Cutting edge: a cis-acting DNA element targets AID-mediated sequence diversification to the chicken Ig light chain gene locus. *J Immunol* **180**: 2019–2023.
41. Kouzarides T. (2007). Chromatin modifications and their function. *Cell* **128**: 693–705.
42. Kuang F.L., Luo Z., Scharff M.D. (2009). H3 trimethyl K9 and H3 acetyl K9 chromatin modifications are associated with class switch recombination. *Proc Natl Acad Sci USA* **106**: 5288–5293.
43. Kuppers R., Dalla-Favera R. (2001). Mechanisms of chromosomal translocations in B cell lymphomas. *Oncogene* **20**: 5580–5594.
44. Kuzin, Ii, Ugine G.D., Wu D. *et al.* (2000). Normal isotype switching in B cells lacking the I mu exon splice donor site: evidence for multiple I mu-like germline transcripts. *J Immunol* **164**: 1451–1457.

45. Lamond A.I., Spector D.L. (2003). Nuclear speckles: a model for nuclear organelles. *Nat Rev Mol Cell Biol* **4**: 605–612.
46. Lebecque S.G., Gearhart P.J. (1990). Boundaries of somatic mutation in rearranged immunoglobulin genes: 5' boundary is near the promoter, and 3' boundary is approximately 1 kb from V(D)J gene. *J Exp Med* **172**: 1717–1727.
47. Li B., Carey M., Workman J.L. (2007). The role of chromatin during transcription. *Cell* **128**: 707–719.
48. Li Z., Luo Z., Scharff M.D. (2004). Differential regulation of histone acetylation and generation of mutations in switch regions is associated with Ig class switching. *Proc Natl Acad Sci USA* **101**: 15428–15433.
49. Lieb J.D., Clarke N.D. (2005). Control of transcription through intragenic patterns of nucleosome composition. *Cell* **123**: 1187–1190.
50. Liu M., Duke J.L., Richter D.J. *et al.* (2008). Two levels of protection for the B cell genome during somatic hypermutation. *Nature* **451**: 841–845.
51. Lorenz M., Jung S., Radbruch A. (1995). Switch transcripts in immunoglobulin class switching. *Science* **267**: 1825–1828.
52. Machida K., Cheng K.T., Pavio N. *et al.* (2005). Hepatitis C virus E2-CD81 interaction induces hypermutation of the immunoglobulin gene in B cells. *J Virol* **79**: 8079–8089.
53. Majumder P., Gomez J.A., Chadwick B.P. *et al.* (2008). The insulator factor CTCF controls MHC class II gene expression and is required for the formation of long-distance chromatin interactions. *J Exp Med* **205**: 785–798.
54. Manis J.P., Tian M., Alt F.W. (2002). Mechanism and control of class-switch recombination. *Trends Immunol* **23**: 31–39.
55. Manis J.P., Van Der Stoep N., Tian M. *et al.* (1998). Class switching in B cells lacking 3' immunoglobulin heavy chain enhancers. *J Exp Med* **188**: 1421–1431.
56. Martin A., Bardwell P.D., Woo C.J. *et al.* (2002). Activation-induced cytidine deaminase turns on somatic hypermutation in hybridomas. *Nature* **415**: 802–806.
57. Martin A., Scharff M.D. (2002a). AID and mismatch repair in antibody diversification. *Nat Rev Immunol* **2**: 605–614.
58. Martin A., Scharff M.D. (2002b). Somatic hypermutation of the AID transgene in B and non-B cells. *Proc Natl Acad Sci USA* **99**: 12304–12308.
59. Martomo S.A., Gearhart P.J. (2006). Somatic hypermutation: subverted DNA repair. *Curr Opin Immunol.* **18**: 243–248.
60. Masuda K., Ouchida R., Takeuchi A. *et al.* (2005). DNA polymerase theta contributes to the generation of C/G mutations during somatic hypermutation of Ig genes. *Proc Natl Acad Sci USA* **102**: 13986–13991.
61. Matsumoto Y., Marusawa H., Kinoshita K. *et al.* (2007). Helicobacter pylori infection triggers aberrant expression of activation-induced cytidine deaminase in gastric epithelium. *Nat Med* **13**: 470–476.
62. McCracken S., Fong N., Rosonina E. *et al.* (1997). 5'-Capping enzymes are targeted to pre-mRNA by binding to the phosphorylated carboxy-terminal domain of RNA polymerase II. *Genes Dev* **11**: 3306–3318.
63. Miele A., Dekker J. (2009). Mapping cis- and trans-chromatin interaction networks using chromosome conformation capture (3C). *Methods Mol Biol* **464**: 105–121.

64. Muramatsu M., Kinoshita K., Fagarasan S. *et al.* (2000). Class switch recombination and hypermutation require activation-induced cytidine deaminase (AID), a potential RNA editing enzyme. *Cell* **102**: 553–563.
65. Murrell A., Heeson S., Reik W. (2004). Interaction between differentially methylated regions partitions the imprinted genes Igf2 and H19 into parent-specific chromatin loops. *Nat Genet* **36**: 889–893.
66. Nambu Y., Sugai M., Gonda H. *et al.* (2003). Transcription-coupled events associating with immunoglobulin switch region chromatin. *Science* **302**: 2137–2140.
67. Neuberger M.S., Di Noia J.M., Beale R.C. *et al.* (2005). Somatic hypermutation at A.T pairs: polymerase error versus dUTP incorporation. *Nat Rev Immunol* **5**: 171–178.
68. Okazaki I.M., Hiai H., Kakazu N. *et al.* (2003). Constitutive expression of AID leads to tumorigenesis. *J Exp Med* **197**: 1173–1181.
69. Osborne C.S., Chakalova L., Brown K.E. *et al.* (2004). Active genes dynamically colocalize to shared sites of ongoing transcription. *Nat Genet* **36**: 1065–1071.
70. Osborne C.S., Chakalova L., Mitchell J.A. *et al.* (2007). Myc dynamically and preferentially relocates to a transcription factory occupied by Igh. *PLoS Biol* **5**: e192.
71. Pandit S., Wang D., Fu X.D. (2008). Functional integration of transcriptional and RNA processing machineries. *Curr Opin Cell Biol* **20**: 260–265.
72. Pasqualucci L., Neumeister P., Goossens T. *et al.* (2001). Hypermutation of multiple proto-oncogenes in B-cell diffuse large-cell lymphomas. *Nature* **412**: 341–346.
73. Peled J.U., Kuang F.L., Iglesias-Ussel M.D. *et al.* (2008). The biochemistry of somatic hypermutation. *Annu Rev Immunol* **26**: 481–511.
74. Perlot T., Alt F.W. (2008). Cis-regulatory elements and epigenetic changes control genomic rearrangements of the IgH locus. *Adv Immunol* **99**: 1–32.
75. Peters A., Storb U. (1996). Somatic hypermutation of immunoglobulin genes is linked to transcription initiation. *Immunity* **4**: 57–65.
76. Pham P., Bransteitter R., Petruska J. *et al.* (2003). Processive AID-catalysed cytosine deamination on single-stranded DNA simulates somatic hypermutation. *Nature* **424**: 103–107.
77. Phillips J.E., Corces V.G. (2009). CTCF: master weaver of the genome. *Cell* **137**: 1194–1211.
78. Pinaud E., Khamlichi A.A., Le Morvan C. *et al.* (2001). Localization of the 3' IgH locus elements that effect long-distance regulation of class switch recombination. *Immunity* **15**: 187–199.
79. Pokholok D.K., Harbison C.T., Levine S. *et al.* (2005). Genome-wide map of nucleosome acetylation and methylation in yeast. *Cell* **122**: 517–527.
80. Rada C., Yelamos J., Dean W. *et al.* (1997). The 5' hypermutation boundary of kappa chains is independent of local and neighbouring sequences and related to the distance from the initiation of transcription. *Eur J Immunol* **27**: 3115–3120.

81. Rajagopal D., Maul R.W., Ghosh A. *et al.* (2009). Immunoglobulin switch mu sequence causes RNA polymerase II accumulation and reduces dA hypermutation. *J Exp Med* **206**: 1237–1244.
82. Ramiro A.R., Stavropoulos P., Jankovic M. *et al.* (2003). Transcription enhances AID-mediated cytidine deamination by exposing single-stranded DNA on the nontemplate strand. *Nat Immunol* **4**: 452–456.
83. Rosenberg B.R., Papavasiliou F.N. (2007). Beyond SHM and CSR: AID and related cytidine deaminases in the host response to viral infection. *Adv Immunol* **94**: 215–244.
84. Rougemaille M., Dieppois G., Kisseleva-Romanova E. *et al.* (2008). THO/Sub2p functions to coordinate 3'-end processing with gene-nuclear pore association. *Cell* **135**: 308–321.
85. Saribasak H., Rajagopal D., Maul R.W. *et al.* (2009). Hijacked DNA repair proteins and unchained DNA polymerases. *Philos Trans R Soc Lond B Biol Sci* **364**: 605–611.
86. Saunders A., Core L.J., Lis J.T. (2006). Breaking barriers to transcription elongation. *Nat Rev Mol Cell Biol* **7**: 557–567.
87. Sekimata M., Perez-Melgosa M., Miller S.A. *et al.* (2009). CCCTC-binding factor and the transcription factor T-bet orchestrate T helper 1 cell-specific structure and function at the interferon-gamma locus. *Immunity* **31**: 551–564.
88. Sellars M., Reina-San-Martin B., Kastner P. *et al.* (2009). Ikaros controls isotype selection during immunoglobulin class switch recombination. *J Exp Med* **206**: 1073–1087.
89. Shahbazian M.D., Grunstein M. (2007). Functions of site-specific histone acetylation and deacetylation. *Annu Rev Biochem* **76**: 75–100.
90. Shen H.M., Storb U. (2004). Activation-induced cytidine deaminase (AID) can target both DNA strands when the DNA is supercoiled. *Proc Natl Acad Sci USA* **101**: 12997–13002.
91. Spilianakis C.G., Flavell R.A. (2004). Long-range intrachromosomal interactions in the T helper type 2 cytokine locus. *Nat Immunol* **5**: 1017–1027.
92. Stavnezer J. (2000). Molecular processes that regulate class switching. *Curr Top Microbiol Immunol* **245**: 127–168.
93. Stavnezer J., Amemiya C.T. (2004). Evolution of isotype switching. *Semin Immunol* **16**: 257–275.
94. Stavnezer J., Guikema J.E., Schrader C.E. (2008). Mechanism and regulation of class switch recombination. *Annu Rev Immunol* **26**: 261–292.
95. Taddei A., Van Houwe G., Hediger F. *et al.* (2006). Nuclear pore association confers optimal expression levels for an inducible yeast gene. *Nature* **441**: 774–778.
96. Takizawa M., Tolarova H., Li Z. *et al.* (2008). AID expression levels determine the extent of cMyc oncogenic translocations and the incidence of B cell tumor development. *J Exp Med* **205**: 1949–1957.
97. Taverna S.D., Li H., Ruthenburg A.J. *et al.* (2007). How chromatin-binding modules interpret histone modifications: lessons from professional pocket pickers. *Nat Struct Mol Biol* **14**: 1025–1040.

98. Teng G., Papavasiliou F.N. (2007). Immunoglobulin somatic hypermutation. *Annu Rev Genet* **41**: 107–120.
99. Tolhuis B., Palstra R.J., Splinter E. *et al.* (2002). Looping and interaction between hypersensitive sites in the active beta-globin locus. *Mol Cell* **10**: 1453–1465.
100. Tornaletti S., Park-Snyder S., Hanawalt P.C. (2008). G4–forming sequences in the non-transcribed DNA strand pose blocks to T7 RNA polymerase and mammalian RNA polymerase II. *J Biol Chem* **283**: 12756–12762.
101. Tous C., Aguilera A. (2007). Impairment of transcription elongation by R-loops in vitro. *Biochem Biophys Res Commun* **360**: 428–432.
102. Tran E.J., Wente S.R. (2006). Dynamic nuclear pore complexes: life on the edge. *Cell* **125**: 1041–1053.
103. Tumas-Brundage K., Manser T. (1997). The transcriptional promoter regulates hypermutation of the antibody heavy chain locus. *J Exp Med* **185**: 239–250.
104. Vakoc C.R., Mandat S.A., Olenchock B.A. *et al.* (2005). Histone H3 lysine 9 methylation and HP1gamma are associated with transcription elongation through mammalian chromatin. *Mol Cell* **19**: 381–391.
105. Vakoc C.R., Sachdeva M.M., Wang H. *et al.* (2006). Profile of histone lysine methylation across transcribed mammalian chromatin. *Mol Cell Biol* **26**: 9185–9195.
106. Wang L., Whang N., Wuerffel R. *et al.* (2006). AID-dependent histone acetylation is detected in immunoglobulin S regions. *J Exp Med* **203**: 215–226.
107. Wang L., Wuerffel R., Feldman S. *et al.* (2009). S region sequence, RNA polymerase II, and histone modifications create chromatin accessibility during class switch recombination. *J Exp Med* **206**: 1817–1830.
108. Woodcock C.L., Dimitrov S. (2001). Higher-order structure of chromatin and chromosomes. *Curr Opin Genet Dev* **11**: 130–135.
109. Wuerffel R., Wang L., Grigera F. *et al.* (2007). S-S synapsis during class switch recombination is promoted by distantly located transcriptional elements and activation-induced deaminase. *Immunity* **27**: 711–722.
110. Xue K., Rada C., Neuberger M.S. (2006). The in vivo pattern of AID targeting to immunoglobulin switch regions deduced from mutation spectra in msh2-/- ung-/- mice. *J Exp Med* **203**: 2085–2094.
111. Yang S.Y., Fugmann S.D., Schatz D.G. (2006). Control of gene conversion and somatic hypermutation by immunoglobulin promoter and enhancer sequences. *J Exp Med* **203**: 2919–2928.
112. Yoshikawa K., Okazaki I.M., Eto T. *et al.* (2002). AID enzyme-induced hypermutation in an actively transcribed gene in fibroblasts. *Science* **296**: 2033–2036.
113. Yu K., Chedin F., Hsieh C.L. *et al.* (2003). R-loops at immunoglobulin class switch regions in the chromosomes of stimulated B cells. *Nat Immunol* **4**: 442–451.
114. Zhao R., Bodnar M.S., Spector D.L. (2009). Nuclear neighborhoods and gene expression. *Curr Opin Genet Dev* **19**: 172–179.

Chapter 3

Cis-Regulatory Elements that Target AID to Immunoglobulin Loci

Sebastian D. Fugmann[1] *and Wesley A. Dunnick*[2]

[1]*Laboratory of Cellular and Molecular Immunology*
National Institute on Aging / National Institutes of Health
Baltimore, MD 21231, USA
E-mail: fugmanns@mail.nih.gov

[2]*Department of Microbiology and Immunolog*
University of Michigan Medical School
Ann Arbor, MI 48103, USA
E-mail: wesadunn@umich.edu

Somatic hypermutation and class switch recombination are regulated genome modification processes that are essential for the diversification and optimization of the immunoglobulin repertoire of B cells after antigen encounter. Both are initiated by the action of activation-induced cytidine deaminase (AID), a powerful mutator enzyme that converts C to U in single-stranded DNA, generated transiently during transcription. *Cis*-regulatory elements play a major role in restricting this mutagenic activity to Ig loci, thereby preventing deleterious effects from occurring in a genome-wide fashion. Recent progress using two robust model systems, the chicken DT40 B cell line and BAC transgenic mice, led to the first identification of such sequence modules. These elements are closely linked to, and may in part overlap with, transcriptional enhancers. The molecular basis of targeting AID-mediated sequence diversification to Ig loci remains to be elucidated, and studies identifying and characterizing distinct important binding sites are only beginning to emerge.

3.1 Introduction

Somatic hypermutation (SHM) is a mutagenic process that randomly alters the DNA sequence of the immunoglobulin (Ig) genes by introducing point mutations and single nucleotide deletions and insertions. The mutagenic activity is focused on the variable (V) regions (referred to in this context as the rearranged VJ or VDJ exon) of the Ig gene loci that encode the antigen-binding sites of the Ig. This allows for the generation of Ig molecules with slightly altered affinity for the cognate antigen, and ultimately, in the context of a germinal center immune response, for the emergence of B cells with higher affinity for the antigen. Class switch recombination (CSR) is a somatic DNA recombination in the Ig heavy chain (IgH) locus that leaves the antigen-binding site intact, but replaces the initial Cμ constant (C) region with another chosen from a set of distinct C regions residing in this locus. As the C regions encode the effector function of the Ig molecule, CSR allows for an optimal immune response based on the class of pathogen the B cell encounters. While the mutagenic activities of both SHM and CSR are highly beneficial for the organism when they alter Ig gene sequences in the context of an immune response, they can, on the other hand, become extremely dangerous when acting genome-wide in an uncontrolled fashion. In such instances, AID-mediated mutagenesis/translocations could lead to the activation of oncogenes and the inactivation of tumor-suppressor genes (Liu and Schatz, 2009). Importantly, such genomic instability/mutator phenotypes are a hallmark of cancers (Hanahan and Weinberg, 2000).

Early sequencing studies indicated that Ig genes are the exclusive target of SHM, and it was thought that this specificity is conferred by distinct *cis*-acting DNA sequences, named "targeting elements". Later on, however, mutations were found in several non-Ig genes (in particular, proto-oncogenes like Bcl-6) in the genome of B cell lymphoma cell lines and primary lymphomas, and the mutation patterns were consistent with that of *bona fide* SHM (Odegard and Schatz, 2006). Such cases, in which SHM hits non-Ig genes is hereafter referred to as "mistargeting". Note that in the context of this chapter, we refer to "targeting" and "targeting elements" as the mechanism and DNA sequences that are responsible for high levels of Ig gene SHM (and CSR), and not to the molecular basis of very low, but still AID-dependent, mutation levels that may occur genome-wide. These low-frequency, but non-Ig restricted mutations are a common challenge when interpreting the findings of many studies on targeting.

CSR is distinguished from SHM by the unique nature of the DNA sequence that represents the substrate for AID, and by the distinct outcome of the reaction, the deletion of large DNA fragments from the Ig locus versus point mutations. Even though the two processes are likely to differ in some aspects of their targeting, they share two potential mechanisms for Ig locus specific-targeting: recruitment of AID, or recruitment of distinct auxiliary factors (including DNA repair enzymes and proteins mediating synapsis) determining the outcome of the reactions. These two potential mechanisms are not mutually exclusive. In the context of this chapter, we will not distinguish between these alternatives and simply refer to targeting and targeting elements, unless explicitly specified.

In the following sections we will discuss the role of promoters and the transcription process itself for targeting and discuss the potential for genome-wide SHM. Subsequently, we will review what is known about *cis*-acting targeting elements for SHM in the murine Ig gene loci, and highlight the recent success of using chicken DT40 cells as a tool towards identifying such elements in the chicken Ig light chain (IgL) gene locus. The next section focuses on what is known about the targeting of CSR and SHM to the murine IgH gene locus. Lastly, we will highlight a few important open questions whose answers might improve our understanding of targeting AID-mediated sequence diversification to Ig gene loci.

3.2 Targeting by Ig Promoters – Are High Levels of Transcription All There is to It?

In vitro, AID deaminates cytosines within single-stranded DNA, but transcription of a double-stranded DNA molecule renders its cytosines accessible as AID substrates (Chaudhuri *et al.*, 2003; Besmer *et al.*, 2006; Sohail *et al.*, 2003; Shen *et al.*, 2009; Bransteitter *et al.*, 2004). It is thought that the unpairing of the DNA strands in the transcription bubble created by the DNA polymerase is responsible for this effect. This requirement for transcription holds true when ectopically-expressing AID in *E. coli* and yeast, and assessing the resulting mutator phenotypes (Sohail *et al.*, 2003; Ramiro *et al.*, 2003; Petersen-Mahrt *et al.*, 2002; Poltoratsky *et al.*, 2004). These findings are consistent with the fact that B cells, by definition, transcribe their rearranged Ig genes and express/secrete the encoded Ig proteins. SHM is strictly dependent on Ig gene transcription (Fukita *et al.*, 1998), and CSR requires the generation of sterile (non-coding) transcripts driven by promoters upstream of switch repeat regions (Stavnezer, 2000). Furthermore, short single-stranded regions are found in V region DNA, coincident with transcription in B cells (Ronai *et al.*, 2007). These requirements for transcription, and the observation that Ig genes are transcribed at particularly high levels in B cells, led to the emergence of the idea that high levels of transcription (and not *cis*-regulatory elements) might be all that is required for targeting of AID-mediated sequence diversification to a given gene. This model is consistent with the observations of AID-dependent mutagenesis in various stably integrated expression cassettes lacking any Ig gene sequences in hybridoma lines, Ramos cells, fibroblasts and CHO cells (Martin and Scharff, 2002; Martin *et al.*, 2002; Yoshikawa *et al.*, 2002). It is important to note, however, that these effects required the ectopic overexpression of AID in these cells, suggesting that high levels of AID might override the action of *cis*-acting targeting elements. Similarly, in non-B cells the targeting machinery might simply be absent or inactive which would also explain why the ubiquitous expression of AID in transgenic mice led to non-B cell tumors without evidence for increased SHM rates in B cells (Okazaki *et al.*, 2003).

3.2.1 *Genome-wide SHM*

One of the strongest challenges to the model of specific *cis*-acting targeting elements came from a study performed by the Wabl laboratory that used a retrovirus containing a GFP reporter gene to test the mutability of numerous randomly selected gene loci (Wang *et al.*, 2004). Surprisingly, the reporter got mutated, as measured by the emergence of GFP^+ cells, in an AID-dependent fashion but independent of where in the genome the reporter was integrated. As the transcription of the reporter gene was driven by the viral LTR promoter, it is consistent with the concept that transcription is all that is required for SHM to occur. A challenge to this interpretation is the very low levels of mutations in this system compared to SHM in activated B cells. Consistent with the low level of mutation, neither of the B cell lines (70Z/3 and 18-81) is of germinal center origin and hence might lack the active targeting machinery. Lastly, a large-scale study in which 118 expressed genes were sequenced from Peyer's patch B cells revealed that one-quarter of genes analyzed was susceptible to SHM (Liu *et al.*, 2008). Importantly, the mutation loads were again very low compared to those found in the Ig genes: in some of the genes mutations could only be detected when DNA repair pathways were disrupted. Overall, while all these reports strongly suggest that non-Ig genes can be targets of SHM, it remains true that the high levels of AID-mediated sequence diversification observed in *bone fide* SHM of V region genes, are exclusive to Ig loci. It is conceivable, however, that genes mutated by mistargeting, contain "attenuated" or partial targeting elements, i.e. only some of the important binding sites are present, or their order/orientation is suboptimal. The identification of the molecular mechanism of targeting will be critical to address this hypothesis.

3.2.2 *Targeting of SHM by promoters*

While it is widely accepted that a transcriptional promoter is essential for all AID-mediated sequence diversification processes, the question whether the very nature of the promoter is important remains open. Some studies show that any promoter is sufficient, but others observe that the ability to drive transcription and the permissiveness for SHM do not correlate. The first notable observation was made in knock-out/knock-in mice in which the endogenous VH promoter (which is PolII-dependent) was either deleted or replaced with a PolI-dependent promoter (Fukita *et al.*, 1998). While the presence of a promoter was required for SHM, even the PolI promoter was unexpectedly able to support it. There was some indication, however, that a cryptic PolII-dependent promoter upstream of the rearranged IgH gene was driving the

SHM process in this instance. Similarly, replacing the Igκ promoter in Igκ transgenes with a heterologous human β-globin promoter, did not affect SHM of the transgene (Betz *et al.*, 1994). But given concerns about the robustness of this model system (discussed below), the relevance of this finding for the endogenous Ig gene loci is unclear. Recent promoter-swapping experiments in the chicken DT40 B cell line (this system is discussed in Section 3.2) revealed that the nature of the promoter itself can drastically alter the frequency of AID-dependent sequence diversification (Yang *et al.*, 2006). When the constitutive EF1α and β-actin promoters were used to replace the endogenous IgL promoter, both were capable of driving high levels of transcription but only the β-actin promoter supported relatively normal levels of AID-mediated sequence diversification. Similar observations were made when the EF1α promoter was replacing the IgL promoter in randomly-integrated reporter constructs containing targeting elements (Kim and Tian, 2009). In contrast, the very same EF1α-driven reporter (and a matching β-actin promoter driven reporter) was getting mutated when integrated in the IgL locus (Kim and Tian, 2009). Note that in this case the endogenous IgL promoter was left intact and might have contributed to targeting. Lastly, the strong Rous Sarcoma Virus (RSV) promoter was sufficient to support SHM of a GFP reporter gene even in the absence of the endogenous IgL promoter (Blagodatski *et al.*, 2009). Overall, these reports suggest that Ig promoters do play an important role in targeting SHM beyond solely driving transcription of the Ig genes. Although Ig promoters are not essential, they do increase the frequency of SHM. Such function is consistent with the model that looping brings *cis*-acting targeting elements in close proximity to the promoters, facilitating the recruitment of AID and/or the transfer of the AID to the transcription machinery that is poised to traverse along the Ig gene (Fig. 3.1). The subsequent movement of the PolII complex along the gene is thought to deliver AID to its cognate DNA substrate, the coding region of the Ig genes for SHM and the switch region for CSR, respectively. Some non-Ig promoters might be permissive for functional interactions with the *cis*-acting targeting elements and hence SHM (or CSR), and some might lack binding sites for *trans*-acting factors critical for looping, and thus do not drive these processes efficiently.

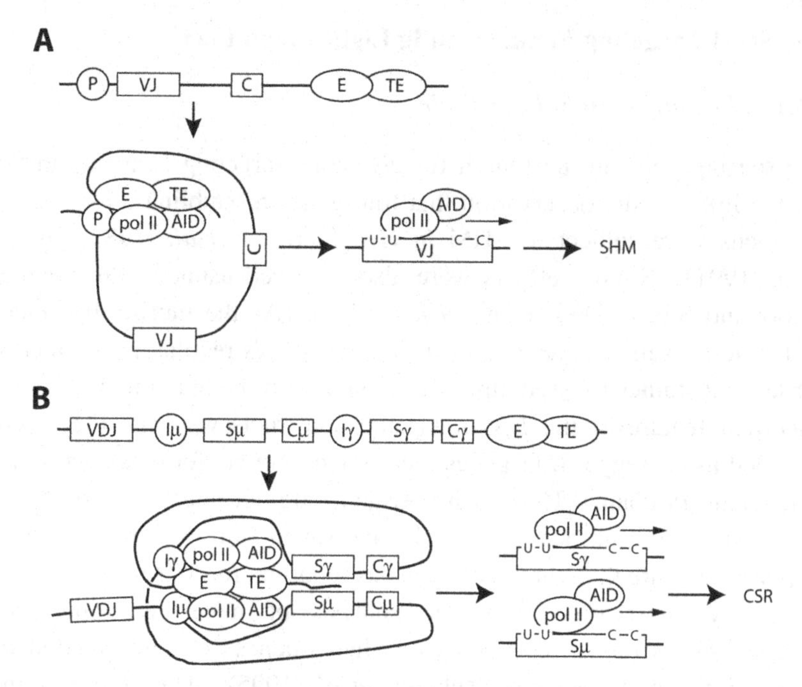

Figure 3.1. **Targeting of SHM and CSR.** A schematic representation of the targeting element model for restricting AID action to Ig gene loci is shown. **(A)** SHM, here depicted for an Ig light chain locus, requires the concerted interaction of the transcriptional enhancer (E) and the targeting element (TE) with the Ig promoter (P) to establish a platform to which RNA polymerase II holoenzyme complexes and AID are recruited. AID then trails with the polymerase complex along the Ig gene and converts Cs to Us in the transiently-formed single-stranded DNA within the VJ exon. The U:G mismatches are subsequently fixed by direct replication and/or error-prone DNA repair to give rise to the SHM end product, a muted V region. **(B)** CSR requires the concerted interaction of the transcriptional enhancer (E), the targeting element (TE) with the intronic promoters (here schematically shown for Iμ and Iγ), which recruits RNA polymerase II and AID to both promoters, and also induces synapsis of the switch repeat regions (Sμ and Sγ, respectively). A high frequency of DNA deamination events within single-stranded DNA formed by transcription of the repeats, leads to DNA double-strand breaks, and ends up with the final CSR product. Note that although transcriptional enhancer and targeting elements are shown as distinct entities, they are likely to show physical overlap in the genome. In addition, the exact number of *cis*-regulatory elements required is unknown, and for simplicity a single element model is shown.

3.3 SHM Targeting Elements in Ig Light Chain Loci

3.3.1 *The murine Ig light chain loci*

The starting point in the search for *cis*-acting targeting elements in the murine Igκ was the observation that transgenes resembling a rearranged Igκ locus were subject to SHM *in vivo* (O'Brien *et al.*, 1987; Sharpe *et al.*, 1991). Similar effects were also observed using Igλ transgenes (Klotz and Storb, 1996; Kong *et al.*, 1998). As the murine Igλ locus is far less well characterized in terms of *cis*-regulatory elements, subsequent studies focused almost exclusively on the Igκ gene locus. An important feature of the Igκ transgenic constructs was that they were designed as passenger transgenes, i.e. they do not confer expression of a functional Igκ chain. Thus such transgenes do not interfere with B cell development, are not subject to selective forces during germinal center responses and are thought to solely act as reporters for SHM.

A key observation from the analysis of mice harboring such transgenes was that heterologous non-Ig sequences can be mutated by SHM within the transgenes (Yelamos *et al.*, 1995). This is in striking contrast to CSR where the nature of the sequence on which AID acts is an essential component for the process to occur (Zarrin *et al.*, 2004). Thus, the search for *cis*-acting targeting elements focused on the non-coding sequences within the Igκ locus. As transcription and SHM are intimately linked, the importance of two well-characterized transcriptional control elements for SHM was tested. Transgenes lacking the intronic κ enhancer (iEκ; located between Jκ5 and Cκ), and its flanking matrix attachment region (MAR) showed no detectable SHM, while transcription was unaffected (Betz *et al.*, 1994). In contrast, Igκ transgenes lacking the 3′ κ enhancer (3′κE) showed a dramatic reduction in both transcription and SHM (Betz *et al.*, 1994). These observations suggested that *cis*-acting targeting elements reside in the iEκ/MAR, while the evidence for such sequences in the 3′κE was inconclusive. Subsequent attempts using this model system to identify the minimal targeting elements for SHM within the iEκ/MAR and the 3′κE by systematic deletion strategies turned out to reveal uninformative results (Goyenechea *et al.*, 1997; Klix *et al.*, 1998). The problems were likely to be of a technical nature, as the behavior of transgenic constructs is frequently strongly dependent on their integration site. In particular, deleting seemingly non-functional "junk" DNA exposes minimal regulatory elements to the influence of their chromosomal environment, and can lead to variegated results. There is anecdotal evidence that a subset of Igκ transgenes containing the full set of regulatory elements do not undergo SHM while still being highly transcribed.

Gene-targeting approaches were used, to avoid the intrinsic challenges encountered when using transgenic constructs to study *cis*-regulatory elements. In contrast to the observations in transgenes, the iEκ element was found to be dispensable for SHM when it was deleted from the endogenous Igκ locus (Inlay *et al.*, 2006). Similarly, the 3′κE enhancer was not essential for SHM, but its deletion led to reduced mutation frequencies that are in part explained by reduced Igκ transcript levels (van der Stoep *et al.*, 1998; Inlay *et al.*, 2006). Lastly, the deletion of the distal enhancer (Ed) led to a similar reduction of both SHM and transcription (Xiang and Garrard, 2008). These findings raise the possibility of redundancy between these transcriptional control elements with respect to SHM targeting. The combined deletion of iEκ and 3′κE results in an early block of B cell development caused by a lack of transcription and hence V(D)J recombination (Inlay *et al.*, 2002), and thus the question of redundancy still awaits to be addressed. Two alternative approaches emerge to bypass the importance of the enhancer elements for transcription during B cell development: (1) the knock-in of exogenous, constitutively active non-Ig promoters allowing for transcription even in the absences of Ig enhancers; and (2) the knock-in of exogenous, constitutively-active non-Ig enhancers. The former might suffer from problems discussed above (non-Ig promoters support only reduced levels of SHM), while the latter might not be sufficient to rescue early B cell development, as is the case for the SV40 enhancer in the context of the IgH locus (Kuzin *et al.*, 2008). In this context, it is also worth noting that the question whether the Igκ coding region is truly irrelevant, has not been formally addressed by gene-targeting thus far.

Overall, the studies of transcriptional enhancers within the murine Igκ locus and their role in SHM, strongly suggest that *cis*-acting targeting elements do exist. They are likely to be linked to, but not necessarily identical with, these transcriptional control elements. Furthermore, as these elements are sensitive with respect to their chromosomal environment, it is obvious that gene-targeting strategies or the use of BAC transgenes (being less susceptible to integration site effects, as discussed below for the IgH locus), are the methods of choice to identify these elements.

3.3.2 *The chicken IgL locus*

The chicken DT40 B cell line is currently considered one of the most robust model systems to address the issue of targeting of SHM. Under standard culture conditions, DT40 cells express AID and continuously undergo SHM and Ig gene conversion (GCV) of the IgH and the single avian IgL gene (Arakawa and Buerstedde, 2004). GCV is closely related to SHM, and thought to only differ in the repair phase of the process, using upstream pseudo-V (ψV) genes as sequence donors for homology-based repair, leading to multiple nucleotide changes per event. The mechanisms that govern targeting of SHM and GCV to Ig loci appear identical, as deletion of all ψV genes shifts the cells to exclusively performing SHM (Arakawa *et al.*, 2004), albeit with a mutation pattern distinct from that observed in humans and mice. Thus the terms SHM- and GCV-targeting are used interchangeably within this section. For the purpose of studying targeting elements, DT40 cells have a key advantage over many other cell line models, namely that standard gene-targeting strategies can be used to manipulate their genome (Arakawa and Buerstedde, 2006). Lastly, the availability of the chicken genome sequence set the stage for the analysis of *cis*-acting DNA sequences in the chicken IgL locus. The genomic assembly of the chicken IgH locus is incomplete in its current state, and thus all studies are focused on the IgL locus, which is more closely related to mammalian Igλ than Igκ.

Three groups employed systematic deletion approaches to identify the *cis*-acting sequences that are required for targeting AID-mediated DNA diversification processes to the IgL gene (Kothapalli *et al.*, 2008; Blagodatski *et al.*, 2009; Kim and Tian, 2009). Although all three studies use slightly different strategies, they converge on the same region of the IgL gene locus located downstream of the C_L constant region exons (Fig. 3.2). The first report of a targeting element by the Fugmann laboratory used the endogenous Ig promoter-driven IgL gene (supported by a SV40 enhancer when endogenous enhancer elements were deleted) as a read-out, and deleted an increasing amount of non-coding DNA from the IgL locus (Kothapalli *et al.*, 2008). This led to the identification of the 3′ regulatory region (3′RR) containing a transcriptional enhancer and a targeting element for SHM. Subsequently, the targeting function was assigned to a 1.3kb subfragment (now named mutational enhancer element, MEE) of the 3′RR, that is essential for SHM but dispensable for transcription in an otherwise unaltered IgL gene locus (N. Kothapalli and S.D. Fugmann, unpublished data; Fig. 3.2). The Buerstedde laboratory used a RSV promoter-driven GFP reporter to assess SHM in the context of an IgL locus in which all pseudo-V elements had been deleted, and identified a bipartite targeting element that was named diversification activator (DIVAC; Blagodatski *et al.*, 2009). Importantly, a DNA fragment containing the DIVAC together with the RSV promoter-driven GFP reporter was sufficient to support AID-mediated sequence diversification when inserted in non-Ig genes. Whether the genes in which these constructs were integrated, were also rendered targets of SHM has not been analyzed thus far. Lastly, the Tian laboratory used a puromycin reporter to assess gene conversion activities, and identified a region "A" that is critical for targeting in the IgL locus, and sufficient for targeting when integrated randomly in the genome, albeit at ten-fold lower frequencies than observed in the IgL locus (Kim and Tian, 2009). This study again focused solely on the reporter gene read-out, and did not report the effects on the respective endogenous IgL and non-Ig genes. A consensus region containing an essential SHM targeting element downstream of the previously identified transcriptional enhancer emerges, that is consistent with findings described in all three reports (Fig. 3.2). The second half of the bipartite DIVAC (5′ of the transcriptional enhancer, see Fig. 3.2) identified by the Buerstedde group, was not detected in any of the other studies. This is likely to be due to differences in the experimental approach, and its importance needs to be addressed by deleting this sequences from an otherwise

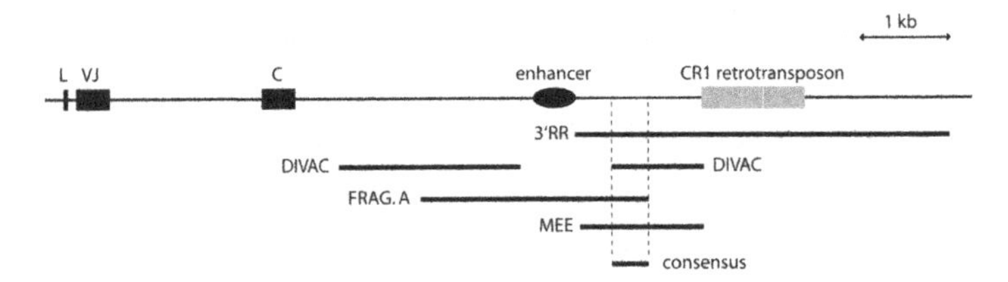

Figure 3.2. **Targeting elements in the chicken IgL locus.** In this schematic representation of a rearranged chicken IgL gene locus the exons are shown as black boxes, the previously reported enhancer as an oval, and the CR1 retrotransposon as a grey box. The regions within the locus that have been shown to be important for targeting of AID-mediated sequence diversification are shown as black lines below the gene. Note that the diversification activator (DIVAC) (Blagodatski *et al.*, 2009) and Frag. A have been identified using reporter genes (Kim and Tian, 2009), while the 3'RR and the mutation enhancer element (MEE) have be found based on their importance for endogenous VJ exon mutations (Kothapalli *et al.*, 2008; N. Kothapalli and S.D. Fugmann, unpublished data). The consensus region corresponds to nucleotides 1147969-1148336 of chromosome 15 in the chicken genome (Gallus gallus assembly 2.1, NW_001471461).

unaltered IgL gene locus. Overall, the DT40 IgL locus model system provides strong support for the existence of *cis*-acting targeting elements, and also suggests that the SHM targeting requires sequence elements distinct from those for enhancing transcription.

Given the size of classic *cis*-regulatory elements (enhancers, silencers and insulators), being in the order of a few hundred base pairs and containing multiple distinct transcription-factor binding sites, it is likely that the targeting elements are of similar size and exhibit similar features. Standard transcription-factor binding site prediction programs reveal a large number of putative binding sites within the IgL MEE (S.D. Fugmann, unpublished data), but the relevance of each of these sites remains to be determined. Interestingly, E2A sites are present in all murine Ig loci and were also found to be enriched in the vicinity of the non-Ig genes undergoing AID-dependent mutagenesis (Liu *et al.*, 2008). E2A sites also act as enhancers of SHM when present in SHM reporter transgenes in mice (Michael *et al.*, 2003). Lastly, a recent report suggested that NFκB, octamer and Mef2 binding sites might be components of the targeting element in the DT40 IgL locus, as altering these sequences led to a significant reduction in gene conversion in the context of an artificial reporter construct (Kim and Tian, 2009).

However, the impact of mutating any such individual binding sites has not been assessed in the context of the endogenous IgL gene locus thus far. On the other hand, DT40 cells deficient in respective transcription factors (c-Rel, p50, E2A) showed reduced frequencies of SHM/GCV (Kim and Tian, 2009; Schoetz *et al.*, 2006), but it is unclear at this point whether they act directly by binding to the IgL locus, or indirectly by controlling the transcription of genes involved in these processes.

3.4 Targeting Elements in the Murine IgH Locus

3.4.1 *Targeting of CSR*

CSR, in its simplest conception, is a DNA deletion that begins in a region between the VDJ exon and the Cμ exon, and ends upstream of one of the Cγ, Cε or Cα genes. The regions in which the deletion begins and ends have been called "S regions" for their role in switch recombination (Fig. 3.3). S regions are composed of simple sequences repeated in tandem over 2–9kb, and lie about 3kb 5′ of each of the C_H gene, except for Cδ. The consequence of the 40–170kb switch deletion is to move the VDJ exon into physical and functional association with a new C_H gene, so that the exon encoding the antigen-binding domain is now associated with a new effector function (i.e. a new C_H region). The ends of the deleted DNA are also ligated to one another, forming a circle (von Schwedler *et al.*, 1990; Iwasato *et al.*, 1990; Matsuoka *et al.*, 1990), and so the class switch is a quasi-reciprocal recombination event. In fact, class switching is most often referred to as a "recombination" event, emphasizing that it involves the ligation of two pieces of DNA.

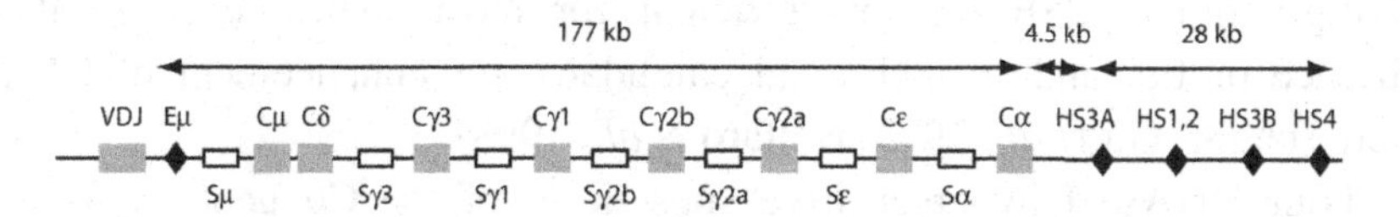

Figure 3.3. **The murine IgH locus.** In this schematic representation of the 3′ end of the murine IgH locus, the rearranged VDJ exon and the constant (C) region exons are shown as grey boxes, the switch (S) regions as open boxes, and the enhancer (E) elements and hypersensitive sites (HS) as black diamonds. The size and distance of important elements is shown above the gene locus.

In order for CSR to occur, AID must deaminate cytosines in S regions (and not in other DNA; Longerich *et al.*, 2006; Chaudhuri and Alt, 2004). Next, repair/recombination enzymes must process the U in DNA to create what is likely to be a double-stranded break (usually with short or long single-stranded flaps; Longerich *et al.*, 2006; Chaudhuri and Alt, 2004). It is also likely that one type of processing is to fill in the single-stranded flaps by error-prone DNA synthesis, resulting in flush double-stranded breaks (Chen *et al.*, 2001; Chaudhuri and Alt, 2004; Longerich *et al.*, 2006). Two double-stranded breaks, one from the Sμ region and one from the Sγ, Sε or Sα region must synapse. Finally, the two breaks must be ligated together if CSR is to actually exchange C_H regions. A goal of many investigators has been to identify the *cis*-acting elements that would help bring all these factors to S regions and would promote S region synapsis.

3.4.1.1 *Enhancers of transcription as enhancers of recombination*

Transcriptional enhancers have been logical candidates for elements that would also improve CSR. Enhancers have a clear role in the regulation of V(D)J recombination (Sleckman *et al.*, 1996). Indeed, the enhancer found between the VDJ exon and Sμ (termed "Eμ") has a measurable, albeit modest, effect on CSR (Gu *et al.*, 1993; Bottaro *et al.*, 1998; Sakai *et al.*, 1999). Normally, CSR is very efficient: virtually all hybridomas expressing IgG, both heavy chain alleles have undergone some DNA rearrangement, and this is usually a deletion between Sμ and Sγ. However, 25% of heavy chain alleles with a deletion of the Eμ region fail to undergo a recombination event, and remain in the germline configuration. CSR remains efficient for most alleles, even in the absence of Eμ, and so additional candidates for enhancement of CSR were sought (Gu *et al.*, 1993; Bottaro *et al.*, 1998).

Four DNAse I hypersensitive sites lie 3′ of the Cα gene in mice, called HS3A, HS1,2, HS3B and HS4 in their order from most Cα proximal to distal (Fig. 3.3; Cogne and Birshtein 2004). HS1,2 represents two DNAse hypersensitive sites that are so close to each other that they are always evaluated as a single regulatory element. HS3A and HS3B are so named because they have almost identical DNA sequences, but these two elements are inverted relative to one another. The four sets of DNAse I hypersensitive sites are spaced over a region from 4 to 32kb 3′ of Cα.

Like most DNAse I hypersensitive sites, the sites 3′ of Cα are regulatory elements. They are enhancers of transcription in classical transient transfection-reporter gene assays (Pettersson *et al.*, 1990; Dariavach *et al.*, 1991; Lieberson *et al.*, 1991; Giannini *et al.*, 1993; Matthias and Baltimore, 1993; Ong *et al.*, 1998), and of stably integrated reporter constructs or heavy chain minigenes (Madisen and Groudine, 1994; Shi and Eckhardt, 2001; Zhang *et al.*, 2007). Importantly, each HS element is a modest enhancer on its own, but three or four of the elements together demonstrate a dramatic synergy for transcriptional enhancement (Lieberson *et al.*, 1991; Chauveau *et al.*, 1998). Furthermore, this enhancer activity is B cell-specific, little or no activity is observed in T cell lines or in fibroblasts (Lieberson *et al.*, 1991; Matthias and Baltimore, 1993; Madisen and Groudine, 1994). B cell-specific, highly synergistic transcriptional enhancement has also been observed in normal B cells by transient transfection (Ong *et al.*, 1998) and in transgenic mice (Chauveau *et al.*, 1999).

Since their discovery, the heavy chain 3′ enhancers have been candidates for enhancers of CSR. One prediction is that they would enhance the germline transcription of γ, ε, α heavy chain genes that precedes CSR (Cogne and Birshtein, 2004). The 3′ HS sites are impressive enhancers of germline transcription (Collins and Dunnick, 1999; Laurencikiene *et al.*, 2007). In addition, germline transcripts of most heavy chain genes are induced by specific cytokines. This cytokine dependence was not observed when all four HS elements, in a 10kb cassette, were used to enhance γ2a germline transcripts; the amount of transcripts was increased 100–fold with or without interferon-γ (Collins and Dunnick, 1999). The combination of two HS sites, however, resulted in IL-4-regulated, high-level transcription of an ε germline gene (Laurencikiene *et al.*, 2007). The combination of these results has multiple valid interpretations. One interpretation is that the enhancer activity of the 3′ HS sites can be revealed by the enhancement provided by two elements in close proximity to the germline promoter. But the physiological enhancement by the 3′ HS sites could be dependent upon their spacing relative to both the promoters they enhance and their spacing relative to one another.

The best test of the 3′ HS sites in germline transcription and CSR would include their deletion from the germline, using homologous recombination in ES cells. Replacement of single HS sites by neomycin expression cassettes has a dramatic effect on germline transcription and CSR for most heavy chain genes (Cogne *et al.*, 1994; Manis *et al.*, 1998; Seidl *et al.*, 1999). In contrast, the "clean" deletion (with the neomycin resistance cassette removed) of any single element results in a modest reduction in CSR at best, with normal expression of all heavy chain genes (Manis *et al.*, 1998; Vincent-Fabert *et al.*, 2009), completely consistent with the synergistic and redundant transcriptional enhancer activity already observed (Ong *et al.*, 1998; Madisen and Groudine, 1994; Chauveau *et al.*, 1998). In light of the results of "clean" deletions, the neomycin replacement cassettes have been interpreted to redirect enhancer activities to their strong *pgk* promoters, away from the germline promoters. Thus, both germline transcription and subsequent CSR are reduced for most heavy chain genes. It is of interest that only genes upstream of the neomycin replacement are affected: the germline transcription and CSR of heavy chain genes between the replacement and the 3′ enhancers remain at wild-type levels (Seidl *et al.*, 1999).

Perhaps due to the reiterated sequences that lie between HS3A, HS1,2 and HS3B, homologous recombination of the 3′ enhancer region in ES cells has proven to be very difficult. Only one clean deletion of multiple HS sites, HS3B and HS4, has been accomplished (Pinaud *et al.*, 2001). This deletion proved that the 3′ enhancers were involved in CSR. Both germline transcription and CSR of the γ3 and γ2b were profoundly reduced, expression of the γ2a, ε and α genes was reduced to about 10% of wild-type levels, but expression of both μ and γ1 was reduced only to 30–50% of wild-type levels (Pinaud *et al.*, 2001). There was a strong coincidence with the neomycin cassette replacement studies as γ1 expression was reduced only slightly, or not at all, by replacement mutations (Cogne *et al.*, 1994; Manis *et al.*, 1998; Seidl *et al.*, 1999). There remained two significant unknowns: (1) would deletion of all four HS sites reduce germline transcription of the ε, α and all four γ heavy chain genes? (2) is each of the four HS sites equivalent, or would deletion of some combinations of HS sites have different phenotypes than deletion of other combinations?

3.4.1.2 *Analysis of IgH 3′ enhancer function by BAC transgenesis*

The Dunnick laboratory has taken a transgenic approach to study of the function of the 3′ enhancer region (Dunnick *et al.*, 2009). A 230kb bacterial artificial chromosome (BAC), including all seven constant region genes, each of the four sets of 3′ HS sites and an additional 15kb of DNA more 3′, was used as the starting point (Fig. 3.3). A VDJ exon expressing anti-arsonate (ARS) binding activity was introduced into this germline BAC at its natural position. The BAC was injected into fertilized mouse eggs, and transgenic mice were produced. Remarkably, the transgenic heavy chain locus undergoes germline transcription and CSR to all heavy chain genes that is regulated like that of the endogenous genes and is quantitatively similar in magnitude (Dunnick *et al.*, 2009).

One of the advantages of the BAC transgenic system is that it is possible to introduce any desired mutation into the BAC heavy chain locus, using homologous recombination in E. coli with inexpensive, but powerful, antibiotic selection (Yang *et al.*, 1997). Two loxP sites were introduced into the BAC just in front of HS3A and just behind HS4, and after producing mice with this intact transgenic heavy chain locus, they were mated to Cre-expressing transgenic mice (Lakso *et al.*, 1996). Thereby, paired transgenic lines were generated that shared the same chromosomal insertion site, retained identical or similar numbers of the transgene but lacked the 3′ enhancer region (Dunnick *et al.*, 2009). In the transgenes lacking the 3′ enhancer region, CSR was reduced for the ε, α and all four γ heavy chain genes to about 1% the level of the intact genes. For the two genes tested (γ1 and γ2a), recombination at the DNA level was undetectable for transgenes lacking the 3′ enhancer region. Therefore, even though the 3′ enhancers may also enhance transcription of the $VDJC_H$ gene after switch recombination (Lieberson *et al.*, 1995), they must enhance the recombination event itself. In the absence of the

3′ enhancers, virtually no switched genes are generated to be transcribed (Dunnick *et al.*, 2009).

For some genes (γ3, γ2b and ε), germline transcription was similarly reduced. For other genes (γ1, γ2a and α), germline transcription was reduced to only 10–50% of the intact transgenes, consistent with the fact that some heavy chain genes have enhancers for germline transcription associated with the heavy chain gene itself (Collins and Dunnick, 1999; Laurencikiene *et al.*, 2007; Xu and Stavnezer, 1992; Lin *et al.*, 1992; Elenich *et al.*, 1996). For these genes, however, germline transcription and CSR could be dissociated. For example, even 30–50% of normal germline transcription could not rescue CSR of the γ1 gene as CSR remained about 1% of the normal level. Thus, CSR to all heavy chain genes depends on the combination of the four HS sites, and the correlation between enhancement of germline transcription and CSR is poor (Dunnick *et al.*, 2009).

Each of the 3′ HS sites occupies less than 1kb of DNA. To test if it is the HS sites/enhancers themselves that enhance CSR, four 1kb deletions were engineered in the heavy chain BAC, and subsequently transgenes were generated lacking only these four kb (W. Dunnick, J. Collins, J. Shi, and C. Fontaine, unpublished data). Tests of the ability of these heavy chain transgenes to undergo germline transcription and CSR will determine if the 24kb of DNA sequence in between the four HS sites have any role in CSR, or if all the enhancement of CSR is encoded in the four HS sites themselves. This experiment is relatively straightforward using the heavy chain BAC: it would be difficult or impossible to target the same heavy chain locus four or more times to produce a similar set of deletions of the endogenous locus.

3.4.2 *Is enhancement of CSR only secondary to enhancement of germline transcription?*

One model for how the 3′ HS sites enhance CSR is that they are merely enhancers of germline transcription. This enhanced germline transcription in turn makes the S regions accessible to AID, perhaps by rendering it single-stranded (Yu *et al.*, 2003). The simplest version of this model cannot be correct. Mutation of a Stat6 binding site in the promoter for germline transcripts of the murine γ1 gene (induced by IL-4) was found to reduce germline transcription, but had no effect on CSR to γ1 (Dunnick *et al.*, 2004). In addition, there are several examples of mutated heavy chain transgenes that reduce germline transcription to 30–50% of the levels for intact loci (usually the γ1 and α genes), but reduce CSR to about 1% of the levels of intact genes (Dunnick *et al.*, 2009 and unpublished). Therefore, for a series of mutations, the amount of CSR is poorly correlated with the amount of germline transcription. The 3′ HS sites do not increase CSR by only increasing the amount of germline transcription. It is more likely that they have a transcription-independent role in attracting AID, "synapse" factors or other repair/recombination factors to the S regions.

3.4.3 *Targeting of SHM to the murine IgH loci*

Initially, IgH passenger transgenes, conceptually identical to those used to delineate Igκ targeting elements, were generated including the intronic Eμ and the 3′ enhancer. Similar to the Igk experiments, the results obtained with this strategy were not robust enough to identify and characterize targeting elements. Some transgenes containing the full set of enhancer elements underwent SHM (Terauchi *et al.*, 2001), while others, although actively transcribed in activated B cells, revealed no detectable levels of SHM (Johnston *et al.*, 1996). The subsequent experiments indicating a role for HS3B and HS4 are thus difficult to interpret (Terauchi *et al.*, 2001). Some of the animal models obtained by BAC transgenesis and gene-targeting described in the previous section of CSR-targeting, have also been analyzed with respect to SHM. The key observations are, that the intronic Eμ enhancer is dispensable for SHM (as it is for CSR) in the context of the endogenous locus (Ronai *et al.*, 2005; Perlot *et al.*, 2005). Similarly, the HS3B/HS4 region of the 3′ IgH is not essential for SHM when deleted in mice (Morvan *et al.*, 2003), contradicting earlier results from transgenic animals (Terauchi *et al.*, 2001). Lastly, BAC transgenics harboring the same 230kb heavy chain transgene described above showed robust frequencies of SHM ($3.6–7.2\times10^{-3}$ per bp; Dunnick *et al.*, 2009). Upon targeted deletion of the entire 3′ enhancer section (including HS3A, HS1,2, HS3B and HS4), the frequencies dropped roughly four-fold to ($0.6–2.3 \times 10^{-3}$ per bp), suggesting a role for HS3A and HS1,2 in SHM targeting. Transcription, however, was also reduced three- to five-fold, raising the question whether this represents the underlying cause of reduced SHM.

One additional surprising observation was that after immunization of these mice, most of the IgH chain mRNA encoded the transgenic VDJ linked to an *endogenous* Cγ gene (Dunnick *et al.*, 2009). CSR within the transgene lacking the 3′ enhancers is so crippled that recombination events with the endogenous locus (even if it is on a different chromosome) are easier to detect (Dunnick *et al.*, 2009; Durdik *et al.*, 1989; Giusti and Manser, 1993). Such recombination between the transgene and the endogenous IgH genes has important technical implications for studies of SHM. When sequencing the transgenic VDJ exon and 3′ flanking regions, it is impossible to know if those sequences remain associated with the transgene, or have recombined, outside the area amplified and sequenced, to the endogenous locus. Taking the heavy chain mRNAs we isolated as representative, one might assume that the majority of the sequences from transgenes with the 3′ enhancer deletion are actually derived from the endogenous locus. In turn, it seems likely that most, or even all, of the mutations Dunnick and colleagues attributed to loci lacking the 3′ enhancers are due to enhancement by the endogenous 3′ regulatory region. It may be possible to test this possibility by breeding transgenic mice that cannot undergo CSR, but retain the ability to somatically mutate their variable regions.

Overall, the studies of SHM-targeting elements in the murine IgH locus provide evidence for the existence of such elements, but their precise location and nature remains elusive. While it seems likely that the 3′ enhancer harbors some of these elements, none of the experimental evidence provided thus far is definitive. It is conceivable that the targeting elements reside outside the known transcriptional control elements, and outside the 230kb of the heavy chain locus BAC. Such putative elements could be hundreds of kb away, either further 3′ of the constant region coding elements, in the V_H cluster, or 5′ of the V_H region cluster. A major obstacle towards identifying these elements is the size of the locus (making an unbiased deletion strategy a challenging task), the lack of reasonable candidate sequences for SHM targeting elements and the potential redundancy of such elements.

3.4.4 *Targeting elements for CSR and SHM – A comparison*

Our knowledge of *cis*-acting targeting elements has rapidly increased over the past few years. A common theme is that targeting elements seem to overlap, physically, with transcriptional enhancers. While it is formally possible that the two types of elements are completely distinct (i.e. not a single *trans*-acting factor is shared), it is more likely that only some parts are distinct, while a larger number of factor binding sites are shared. The transcriptional enhancers do provide the chromatin environment to allow transcription to occur, and are thus also indispensable for SHM and CSR. Interestingly, while high levels of transcription are important for SHM occur, the germline promoters driving the sterile transcripts essential for CSR are not strong promoters at all: the amount of sterile Iγ or Iα transcripts is only a few percent of those coding rearranged VDJCμ transcripts (Stavnezer-Nordgren and Sirlin, 1986; Yancopoulos *et al.*, 1986). This suggests that the relationship between transcription and recruitment of AID is different for SHM and CSR.

The targeting elements themselves have at most a minimal role in the enhancement of transcription, but require a transcriptionally-active environment to exert their function. Their mode of action is unknown, but it is conceivable that they could act by recruiting AID to Ig gene loci, and/or by recruiting DNA recombination/repair factors required to process the initial AID-induced DNA lesion, or in a yet unknown fashion. The key difference between SHM and CSR is the ultimate outcome of the reactions, point mutations and recombination, respectively. This is driven by the distinct sequence AID acts upon, which contains a high density of repeats in the case of CSR but almost no repetitive elements in the case of SHM. For efficient CSR to occur, a close spatial proximity of two switch repeat regions is critical (Chaudhuri *et al.*, 2007), and facilitating such synapsis may represent a unique feature of CSR-targeting elements. In addition, the requirements for DNA repair factors differ between SHM and CSR, and distinct components of their targeting elements could also control this aspect. Lastly, as the promoter driving the VDJCμ transcript remains active in B cells activated to undergo CSR expressing sterile Iμ/γ/ε/α transcripts, the targeting elements for SHM and CSR differ in their preference for the promoters they functionally interact with. Overall, it is in these aspects where the CSR- and SHM-targeting elements are likely to turn out to be quite different. To date, the only experimental evidence for this concept come from the studies in which the HS3B/HS4 sites were deleted in the murine IgH locus, leading to a reduction of CSR while leaving SHM unaffected (Morvan *et al.*, 2003).

3.5 Outlook

Two robust model systems have emerged to address the mechanism of targeting AID-mediated sequence diversification: DT40 cells and IgH BAC transgenes. While the former has the potential to provide informative data with respect to SHM, the latter provides insight into both SHM and CSR. These systems have the potential to define the minimal sequence requirement of the *cis*-acting targeting elements. In turn, this information would enhance other investigations into additional critical issues:

1) What are the *trans*-acting factors that mediate the targeting function?
2) What is the mechanism by which *cis*-regulatory elements and *trans*-acting factors act in concert to mediate targeting, and how are neighboring genes spared from the mutagenic activities?
3) How are neighboring genes spared from targeting?
4) What is the molecular basis of mistargeting, in particular of oncogenes?

As *cis*-regulatory elements consist of sets of binding sites for sequence- or structure-specific DNA binding factors, in particular transcription factors, it is likely that the same is true for SHM/CSR targeting elements. Such binding sites are present in a defined local arrangement to provide an assembly platform for the *trans*-acting factors. For transcriptional control elements it has been shown that the spatial orientation and the distance of such sites is important in some cases, but it is irrelevant in others. There is considerable optimism that the continuation of the systematic deletion/mutation strategies in both model systems will provide us with minimal sets of binding sites required for targeting. Although it is possible that the very same combination of sites will emerge for SHM from the murine IgH and chicken IgL locus, it is likely that only a set of core elements is common between these loci. Therefore it will be important to identify SHM-targeting elements from additional Ig loci from multiple species (including, of course, the murine Igκ and Igλ loci). A complementation approach using the chicken IgL locus as a recipient for putative *cis*-regulatory elements shows promising preliminary results (N. Kothapalli and S.D. Fugmann, unpublished data). It seems highly feasible to use a similar approach in the murine IgH BAC transgene system. Such functional complementation studies will be guided by sophisticated bioinformatic strategies to identify sequence elements of interest in the rapidly increasing number of vertebrate genomes becoming available. Similar approaches are also likely to yield relevant information in the context of CSR. Nevertheless, the gold standard for functional analysis of *cis*-regulatory elements is their mutation/deletion in the endogenous locus of the species of origin. The long-term goal of any analysis of targeting elements should be a test of their function in their physiological location.

The identification of the *trans*-acting factors that mediate the targeting is primarily dependent on defining the respective *cis*-regulatory sequences. Subsequent use of standard biochemical and molecular biology approaches to address DNA-protein interactions will help to reveal their identity. While it is commonly thought that a distinct set of classic transcription factors mediate targeting, none of the numerous transcription factor knock-out mice analyzed thus far shows a strong and specific SHM defect. Hence, novel sequence-specific DNA binding proteins, perhaps lacking transcriptional activation domains, might be the *trans*-acting components of targeting. An alternative approach to determine the *trans*-acting targeting factors is currently being pursued by numerous groups in the field – the identification of AID-interacting proteins. This is based on the model that AID gets specifically recruited to Ig gene loci, and although it is likely to be somewhat true, it has not been formally proven as of yet.

An alternative, but not mutually exclusive model, briefly mentioned above, proposes that the error-prone repair machinery and other cofactors (including those involved in synapsis of the switch regions in the case of CSR) are specifically recruited to Ig gene loci. There is clear experimental evidence that error-free DNA repair copes well with AID-induced DNA damage on some non-Ig genes (Liu *et al.*, 2008), but it is likely that AID-induced DNA lesions in such loci are less frequent compared to those occurring in Ig genes. How this locus-specific recruitment of DNA repair proteins occurs, and/or how DNA repair programs switch from error-free to error-prone in distinct chromosomal locations is unknown. This model also raises the intriguing possibility that the targeting elements are not unique but rather members of a large family of *cis*-regulatory sequences controlling genomic integrity. It is tempting to speculate that "stabilizer" sequences ensure highly efficient error-free repair on essential genes, whereas "destabilizer" sequences provide the molecular basis for genomic instability in chromosomal regions where sequence variation could be beneficial for the fitness of the organism (like the Ig genes in the context of the immune system). Such elements are likely to be conferring distinct chromatin structures and properties to the respective gene loci, and similar to the case of transcription, insulator and boundary elements would block one-dimensional spreading of such features along the DNA fiber.

The protection of neighboring genes is currently an underappreciated question. If targeting elements indeed represent focal points for the recruitment of the diversification machinery, it is critical to restrict their activity to the Ig gene sequences themselves, and to suppress their activity on the genes in their direct vicinity. While some studies have suggested that boundary elements (discussed briefly above) can be found in Ig loci (Seaman, 2005; Sepulveda *et al.*, 2005; Gopal and Fugmann, 2008), it also possible that distinct promoter structures (similar to that of the EF1α promoter which is non-permissive for SHM, as discussed above) of the neighboring genes preclude these genes from becoming targets of AID-mediated sequence diversification.

Lastly, as mentioned above, mistargeting of AID-mediated sequence diversification processes is a key step in the development of B cell lymphomas. It is conceivable that mistargeting simply represents less efficient targeting to non-Ig loci. Attenuated targeting could occur either at the level of AID recruitment, or at the level of the DNA repair machinery and its control. Reduced targeting to other genes may be due to a lack of one or more of the important binding sites that form the minimal SHM/CSR targeting elements. Similarly, the presence of "stabilizer" or "destabilizer" elements, unrelated to those present in Ig loci, might predetermine whether an oncogene or tumor suppressor are more likely to be subjected to AID-mediated sequence diversification, in particular in instances when AID is aberrantly expressed.

3.6 Acknowledgements

W.D. would like to thank his colleagues at the University of Michigan and throughout the world for their stimulating discussions and many kind gifts that were essential to convert our ideas to reality. Work in W.D.'s laboratory was supported by the National Institutes of Health (AI076057) and by support of the Transgenic Core Facility of the University of Michigan Cancer Center (CA46592). This work was in part (to S.D.F.) supported by the Intramural Research Program of the NIH / National Institute on Aging.

3.7 References

1. Arakawa H., Buerstedde J.M. (2004). Immunoglobulin gene conversion: insights from bursal B cells and the DT40 cell line. *Dev Dyn* **229**: 458–464.
2. Arakawa H., Buerstedde J.M. (2006). Dt40 gene disruptions: a how-to for the design and the construction of targeting vectors. *Subcell Biochem* **40**: 1–9.
3. Arakawa H., Saribasak H., Buerstedde J.M. (2004). Activation-induced cytidine deaminase initiates immunoglobulin gene conversion and hypermutation by a common intermediate. *PLoS Biol* **2**: E179.
4. Besmer E., Market E., Papavasiliou F.N. (2006). The transcription elongation complex directs activation-induced cytidine deaminase-mediated DNA deamination. *Mol Cell Biol* **26**: 4378–4385.

5. Betz A.G., Milstein C., Gonzalez-Fernandez A. *et al.* (1994). Elements regulating somatic hypermutation of an immunoglobulin kappa gene: critical role for the intron enhancer/matrix attachment region. *Cell* **77**: 239–248.
6. Blagodatski A., Batrak V., Schmidl S. *et al.* (2009). A cis-acting diversification activator both necessary and sufficient for AID-mediated hypermutation. *PLoS Genet* **5**: e1000332.
7. Bottaro A., Young F., Chen J. *et al.* (1998). Deletion of the IgH intronic enhancer and associated matrix-attachment regions decreases, but does not abolish, class switching at the mu locus. *Int Immunol* **10**: 799–806.
8. Bransteitter R., Pham P., Calabrese P. *et al.* (2004). Biochemical analysis of hypermutational targeting by wild type and mutant activation-induced cytidine deaminase. *J Biol Chem* **279**: 51612–51621.
9. Chaudhuri J., Alt F.W. (2004). Class-switch recombination: interplay of transcription, DNA deamination and DNA repair. *Nat Rev Immunol* **4**: 541–552.
10. Chaudhuri J., Basu U., Zarrin A. *et al.* (2007). Evolution of the immunoglobulin heavy chain class switch recombination mechanism. *Adv Immunol* **94**: 157–214.
11. Chaudhuri J., Tian M., Khuong C. *et al.* (2003). Transcription-targeted DNA deamination by the AID antibody diversification enzyme. *Nature* **422**: 726–730.
12. Chauveau C., Jansson E.A., Muller S. *et al.* (1999). Cutting edge: Ig heavy chain 3' HS1-4 directs correct spatial position-independent expression of a linked transgene to B lineage cells. *J Immunol* **163**: 4637–4641.
13. Chauveau C., Pinaud E., Cogne M. (1998). Synergies between regulatory elements of the immunoglobulin heavy chain locus and its palindromic 3' locus control region. *Eur J Immunol* **28**: 3048–3056.
14. Chen X., Kinoshita K., Honjo T. (2001). Variable deletion and duplication at recombination junction ends: implication for staggered double-strand cleavage in class-switch recombination. *Proc Natl Acad Sci USA* **98**: 13860–13865.
15. Cogne M., Birshtein B.K. (2004). Regulation of class switch recombination in T. Honjo, F.W. Alt & M.S. Neuberger (Eds.) *Molecular Biology of B Cells.* London, Academic Press.
16. Cogne M., Lansford R., Bottaro A. *et al.* (1994). A class switch control region at the 3' end of the immunoglobulin heavy chain locus. *Cell* **77**: 737–747.
17. Collins J.T., Dunnick W.A. (1999). Cutting edge: IFN-gamma regulated germline transcripts are expressed from gamma2a transgenes independently of the heavy chain 3' enhancers. *J Immunol* **163**: 5758–5762.
18. Dariavach P., Williams G.T., Campbell K. *et al.* (1991). The mouse IgH 3'-enhancer. *Eur J Immunol* **21**: 1499–1504.
19. Dunnick W.A., Collins J.T., Shi J. *et al.* (2009). Switch recombination and somatic hypermutation are controlled by the heavy chain 3' enhancer region. *J Exp Med* **206**: 2613–2623.
20. Dunnick W.A., Shi J., Graves K.A. *et al.* (2004). Germline transcription and switch recombination of a transgene containing the entire H chain constant region locus: effect of a mutation in a STAT6 binding site in the gamma 1 promoter. *J Immunol* **173**: 5531–5539.

21. Durdik J., Gerstein R.M., Rath S. *et al.* (1989). Isotype switching by a microinjected mu immunoglobulin heavy chain gene in transgenic mice. *Proc Natl Acad Sci USA* **86**: 2346–2350.
22. Elenich L.A., Ford C.S., Dunnick W.A. (1996). The gamma 1 heavy chain gene includes all of the cis-acting elements necessary for expression of properly regulated germ-line transcripts. *J Immunol* **157**: 176–182.
23. Fukita Y., Jacobs H., Rajewsky K. (1998). Somatic hypermutation in the heavy chain locus correlates with transcription. *Immunity* **9**: 105–114.
24. Giannini S.L., Singh M., Calvo C.F. *et al.* (1993). DNA regions flanking the mouse Ig 3' alpha enhancer are differentially methylated and DNAase I hypersensitive during B cell differentiation. *J Immunol* **150**: 1772–1780.
25. Giusti A.M., Manser T. (1993). Hypermutation is observed only in antibody H chain V region transgenes that have recombined with endogenous immunoglobulin H DNA: implications for the location of cis-acting elements required for somatic mutation. *J Exp Med* **177**: 797–809.
26. Gopal A.R., Fugmann S.D. (2008). AID-mediated diversification within the IgL locus of chicken DT40 cells is restricted to the transcribed IgL gene. *Mol Immunol* **45**: 2062–2068.
27. Goyenechea B., Klix N., Yelamos J. *et al.* (1997). Cells strongly expressing Ig(kappa) transgenes show clonal recruitment of hypermutation: a role for both MAR and the enhancers. *EMBO J* **16**: 3987–3994.
28. Gu H., Zou Y.R., Rajewsky K. (1993). Independent control of immunoglobulin switch recombination at individual switch regions evidenced through Cre-loxP-mediated gene targeting. *Cell* **73**: 1155–1164.
29. Hanahan D., Weinberg R.A. (2000). The hallmarks of cancer. *Cell* **100**: 57–70.
30. Inlay M., Alt F.W., Baltimore D. *et al.* (2002). Essential roles of the kappa light chain intronic enhancer and 3' enhancer in kappa rearrangement and demethylation. *Nat Immunol* **3**: 463–468.
31. Inlay M.A., Gao H.H., Odegard V.H. *et al.* (2006). Roles of the Ig kappa light chain intronic and 3' enhancers in Igk somatic hypermutation. *J Immunol* **177**: 1146–1151.
32. Iwasato T., Shimizu A., Honjo T. *et al.* (1990). Circular DNA is excised by immunoglobulin class switch recombination. *Cell* **62**: 143–149.
33. Johnston J.M., Ihyer S.R., Smith R.S. *et al.* (1996). Analysis of hypermutation in immunoglobulin heavy chain passenger transgenes. *Eur J Immunol* **26**: 1058–1062.
34. Kim Y., Tian M. (2009). NF-kappaB family of transcription factor facilitates gene conversion in chicken B cells. *Mol Immunol* **46**: 3283–3291.
35. Klix N., Jolly C.J., Davies S.L. *et al.* (1998). Multiple sequences from downstream of the J kappa cluster can combine to recruit somatic hypermutation to a heterologous, upstream mutation domain. *Eur J Immunol* **28**: 317–326.
36. Klotz E.L., Storb U. (1996). Somatic hypermutation of a lambda 2 transgene under the control of the lambda enhancer or the heavy chain intron enhancer. *J Immunol* **157**: 4458–4463.

37. Kong Q., Zhao L., Subbaiah S. *et al.* (1998). A lambda 3' enhancer drives active and untemplated somatic hypermutation of a lambda 1 transgene. *J Immunol* **161**: 294–301.
38. Kothapalli N., Norton D.D., Fugmann S.D. (2008). Cutting edge: a cis-acting DNA element targets AID-mediated sequence diversification to the chicken Ig light chain gene locus. *J Immunol* **180**: 2019–2023.
39. Kuzin I.I., Bagaeva L., Young F.M. *et al.* (2008). Requirement for enhancer specificity in immunoglobulin heavy chain locus regulation. *J Immunol* **180**: 7443–7450.
40. Lakso M., Pichel J.G., Gorman J.R. *et al.* (1996). Efficient in vivo manipulation of mouse genomic sequences at the zygote stage. *Proc Natl Acad Sci USA* **93**: 5860–5865.
41. Laurencikiene J., Tamosiunas V., Severinson E. (2007). Regulation of epsilon germline transcription and switch region mutations by IgH locus 3' enhancers in transgenic mice. *Blood* **109**: 159–167.
42. Lieberson R., Giannini S.L., Birshtein B.K. *et al.* (1991). An enhancer at the 3' end of the mouse immunoglobulin heavy chain locus. *Nucleic Acids Res* **19**: 933–937.
43. Lieberson R., Ong J., Shi X. *et al.* (1995). Immunoglobulin gene transcription ceases upon deletion of a distant enhancer. *EMBO J* **14**: 6229–6238.
44. Lin Y.C., Shockett P., Stavnezer J. (1992). Regulation of transcription of the germline immunoglobulin alpha constant region gene. *Curr Top Microbiol Immunol* **182**: 157–165.
45. Liu M., Duke J.L., Richter D.J. *et al.* (2008). Two levels of protection for the B cell genome during somatic hypermutation. *Nature* **451**: 841–845.
46. Liu M., Schatz D.G. (2009). Balancing AID and DNA repair during somatic hypermutation. *Trends Immunol* **30**: 173–181.
47. Longerich S., Basu U., Alt F. *et al.* (2006). AID in somatic hypermutation and class switch recombination. *Curr Opin Immunol* **18**: 164–174.
48. Madisen L., Groudine M. (1994). Identification of a locus control region in the immunoglobulin heavy-chain locus that deregulates c-myc expression in plasmacytoma and Burkitt's lymphoma cells. *Genes Dev* **8**: 2212–2226.
49. Manis J.P., Van Der Stoep N., Tian M. *et al.* (1998). Class switching in B cells lacking 3' immunoglobulin heavy chain enhancers. *J Exp Med* **188**: 1421–1431.
50. Martin A., Bardwell P.D., Woo C.J. *et al.* (2002). Activation-induced cytidine deaminase turns on somatic hypermutation in hybridomas. *Nature* **415**: 802–806.
51. Martin A., Scharff M.D. (2002). Somatic hypermutation of the AID transgene in B and non-B cells. *Proc Natl. Acad Sci USA* **99**: 12304–12308.
52. Matsuoka M., Yoshida K., Maeda T. *et al.* (1990). Switch circular DNA formed in cytokine-treated mouse splenocytes: evidence for intramolecular DNA deletion in immunoglobulin class switching. *Cell* **62**: 135–142.
53. Matthias P., Baltimore D. (1993). The immunoglobulin heavy chain locus contains another B-cell-specific 3' enhancer close to the alpha constant region. *Mol Cell Biol* **13**: 1547–1553.

54. Michael N., Shen H.M., Longerich S. *et al.* (2003). The E box motif CAGGTG enhances somatic hypermutation without enhancing transcription. *Immunity* **19**: 235–242.
55. Morvan C.L., Pinaud E., Decourt C. *et al.* (2003). The immunoglobulin heavy-chain locus hs3b and hs4 3' enhancers are dispensable for VDJ assembly and somatic hypermutation. *Blood* **102**: 1421–1427.
56. O'Brien R.L., Brinster R.L., Storb U. (1987). Somatic hypermutation of an immunoglobulin transgene in kappa transgenic mice. *Nature* **326**: 405–409.
57. Odegard V.H., Schatz D.G. (2006). Targeting of somatic hypermutation. *Nat Rev Immunol* **6**: 573–583.
58. Okazaki I.M., Hiai H., Kakazu N. *et al.* (2003). Constitutive expression of AID leads to tumorigenesis. *J Exp Med* **197**: 1173–1181.
59. Ong J., Stevens S., Roeder R.G. *et al.* (1998). 3' IgH enhancer elements shift synergistic interactions during B cell development. *J Immunol* **160**: 4896–4903.
60. Perlot T., Alt F.W., Bassing C.H. *et al.* (2005). Elucidation of IgH intronic enhancer functions via germ-line deletion. *Proc Natl. Acad. Sci. USA* **102**: 14362–14367.
61. Petersen-Mahrt S.K., Harris R.S., Neuberger M.S. (2002). AID mutates E. coli suggesting a DNA deamination mechanism for antibody diversification. *Nature* **418**: 99–103.
62. Pettersson S., Cook G.P., Bruggemann M. *et al.* (1990). A second B cell-specific enhancer 3' of the immunoglobulin heavy-chain locus. *Nature* **344**: 165–168.
63. Pinaud E., Khamlichi A.A., Le Morvan C. *et al.* (2001). Localization of the 3' IgH locus elements that effect long-distance regulation of class switch recombination. *Immunity* **15**: 187–199.
64. Poltoratsky V.P., Wilson S.H., Kunkel T.A. *et al.* (2004). Recombinogenic phenotype of human activation-induced cytosine deaminase. *J Immunol* **172**: 4308–4313.
65. Ramiro A.R., Stavropoulos P., Jankovic M. *et al.* (2003). Transcription enhances AID-mediated cytidine deamination by exposing single-stranded DNA on the nontemplate strand. *Nat Immunol* **4**: 452–456.
66. Ronai D., Iglesias-Ussel M.D., Fan M. *et al.* (2007). Detection of chromatin-associated single-stranded DNA in regions targeted for somatic hypermutation. *J Exp Med* **204**: 181–190.
67. Ronai D., Iglesias-Ussel M.D., Fan M. *et al.* (2005). Complex regulation of somatic hypermutation by cis-acting sequences in the endogenous IgH gene in hybridoma cells. *Proc Natl Acad Sci USA* **102**: 11829–11834.
68. Sakai E., Bottaro A., Alt F.W. (1999). The Ig heavy chain intronic enhancer core region is necessary and sufficient to promote efficient class switch recombination. *Int Immunol* **11**: 1709–1713.
69. Schoetz U., Cervelli M., Wang Y.D. *et al.* (2006). E2A expression stimulates Ig hypermutation. *J Immunol* **177**: 395–400.
70. Seaman M.N. (2005). Recycle your receptors with retromer. *Trends Cell Biol* **15**: 68–75.

71. Seidl K.J., Manis J.P., Bottaro A. *et al.* (1999). Position-dependent inhibition of class-switch recombination by PGK-neor cassettes inserted into the immunoglobulin heavy chain constant region locus. *Proc Natl Acad Sci USA* **96**: 3000–3005.
72. Sepulveda M.A., Garrett F.E., Price-Whelan A. *et al.* (2005). Comparative analysis of human and mouse 3' Igh regulatory regions identifies distinctive structural features. *Mol Immunol* **42**: 605–615.
73. Sharpe M.J., Milstein C., Jarvis J.M. *et al.* (1991). Somatic hypermutation of immunoglobulin kappa may depend on sequences 3' of C kappa and occurs on passenger transgenes. *EMBO J* **10**: 2139–2145.
74. Shen H.M., Poirier M.G., Allen M.J. *et al.* (2009). The activation-induced cytidine deaminase (AID) efficiently targets DNA in nucleosomes but only during transcription. *J Exp Med* **206**: 1057–1071.
75. Shi X., Eckhardt L.A. (2001). Deletional analyses reveal an essential role for the hs3b/hs4 IgH 3' enhancer pair in an Ig-secreting but not an earlier-stage B cell line. *Int Immunol* **13**: 1003–1012.
76. Sleckman B.P., Gorman J.R., Alt F.W. (1996). Accessibility control of antigen-receptor variable-region gene assembly: role of cis-acting elements. *Annu Rev Immunol* **14**: 459–481.
77. Sohail A., Klapacz J., Samaranayake M. *et al.* (2003). Human activation-induced cytidine deaminase causes transcription-dependent, strand-biased C to U deaminations. *Nucleic Acids Res* **31**: 2990–2994.
78. Stavnezer J. (2000). Molecular processes that regulate class switching. *Curr Top Microbiol Immunol* **245**: 127–168.
79. Stavnezer-Nordgren J., Sirlin S. (1986). Specificity of immunoglobulin heavy chain switch correlates with activity of germline heavy chain genes prior to switching. *EMBO J* **5**: 95–102.
80. Terauchi A., Hayashi K., Kitamura D. *et al.* (2001). A pivotal role for DNase I-sensitive regions 3b and/or 4 in the induction of somatic hypermutation of IgH genes. *J Immunol* **167**: 811–820.
81. van Der Stoep N., Gorman J.R., Alt F.W. (1998). Reevaluation of 3'Ekappa function in stage- and lineage-specific rearrangement and somatic hypermutation. *Immunity* **8**: 743–750.
82. Vincent-Fabert C., Truffinet V., Fiancette R. *et al.* (2009). Ig synthesis and class switching do not require the presence of the hs4 enhancer in the 3' IgH regulatory region. *J Immunol* **182**: 6926–6932.
83. von Schwedler U., Jack H.M., Wabl M. (1990). Circular DNA is a product of the immunoglobulin class switch rearrangement. *Nature* **345**: 452–456.
84. Wang C.L., Harper R.A., Wabl M. (2004). Genome-wide somatic hypermutation. *Proc Natl Acad Sci USA* **101**: 7352–7356.
85. Xiang Y., Garrard W.T. (2008). The Downstream Transcriptional Enhancer, Ed, positively regulates mouse Ig kappa gene expression and somatic hypermutation. *J Immunol* **180**: 6725–6732.

86. Xu M.Z., Stavnezer J. (1992). Regulation of transcription of immunoglobulin germ-line gamma 1 RNA: analysis of the promoter/enhancer. *EMBO J* **11**: 145–155.
87. Yancopoulos G.D., DePinho R.A., Zimmerman K.A. *et al.* (1986). Secondary genomic rearrangement events in pre-B cells: VHDJH replacement by a LINE-1 sequence and directed class switching. *EMBO J* **5**: 3259–3266.
88. Yang S.Y., Fugmann S.D., Schatz D.G. (2006). Control of gene conversion and somatic hypermutation by immunoglobulin promoter and enhancer sequences. *J Exp Med* **203**: 2919–2928.
89. Yang X.W., Model P., Heintz N. (1997). Homologous recombination based modification in Escherichia coli and germline transmission in transgenic mice of a bacterial artificial chromosome. *Nat Biotechnol* **15**: 859–865.
90. Yelamos J., Klix N., Goyenechea B. *et al.* (1995). Targeting of non-Ig sequences in place of the V segment by somatic hypermutation. *Nature* **376**: 225–229.
91. Yoshikawa K., Okazaki I.M., Eto T. *et al.* (2002). AID enzyme-induced hypermutation in an actively transcribed gene in fibroblasts. *Science* **296**: 2033–2036.
92. Yu K., Chedin F., Hsieh C.L. *et al.* (2003). R-loops at immunoglobulin class switch regions in the chromosomes of stimulated B cells. *Nat Immunol* **4**: 442–451.
93. Zarrin A.A., Alt F.W., Chaudhuri J. *et al.* (2004). An evolutionarily conserved target motif for immunoglobulin class-switch recombination. *Nat Immunol* **5**: 1275–1281.
94. Zhang B., Alaie-Petrillo A., Kon M. *et al.* (2007). Transcription of a productively rearranged Ig VDJC alpha does not require the presence of HS4 in the IgH 3' regulatory region. *J Immunol* **178**: 6297–6306.

Chapter 4

Partners in Diversity: The Search for AID Co-Factors

Bernardo Reina-San-Martin[1] *and Jayanta Chaudhuri*[2]

[1]*Department of Cancer Biology*
Institut de Génétique et de Biologie Moléculaire et Cellulaire (IGBMC)
Inserm U964 - CNRS UMR7104 - Université de Strasbourg
67404 Illkirch CEDEX, France
E-mail: reinab@igbmc.fr

[2]*Immunology Program, Memorial Sloan Kettering Cancer Center*
New York, NY 10065, USA
E-mail: chaudhuj@mskcc.org

Activation-induced cytidine deaminase (AID) plays an essential role in the B-cell specific immunoglobulin (Ig) gene diversification reactions of class switch recombination (CSR), somatic hypermutation (SHM) and Ig gene conversion (GCV). It is generally believed that AID employs single-strand DNA as a substrate to convert cytosines to uridines within transcribed regions of immunoglobulin loci. Components of several DNA repair pathways, including base excision repair and mismatch repair, process the deaminated residues to trigger mutation and gene conversion during SHM and GCV, or DNA double-strand breaks during CSR. Proteins of the globally expressed end-joining pathways subsequently ligate the DNA breaks to complete CSR. While mutations in AID severely impair the ability of an individual to mount an effective immune response to pathogens, it is becoming increasingly clear that AID has the potential to introduce DNA lesions at regions beyond the immunoglobulin loci. Such mistargeted AID activity plays a significant role in the ontogeny of mature B cell malignancies. It is therefore critical that AID activity is restricted only to defined regions of the genome. How such restriction is effected is still a big unknown, yet a subject of intense effort from multiple laboratories. It is however clear that AID interacts with multiple proteins and such interactions potentially regulate its intracellular localization, activity and targeting to specific regions of the genome. Here we describe the factors with which AID interacts, and discuss the implications of such interactions in AID function and specificity.

4.1 Introduction and Overview

Antibodies are composed of two pairs of identical immunoglobulin heavy (IgH) and light (IgL) chains. Each chain bears a variable region and a constant region. The combination of IgH and IgL variable regions determines the antigen-recognition specificity of the receptor whereas the IgH constant region determines the antibody isotype. The variable regions of IgH and IgL antibody chains are assembled during development in the bone marrow through V(D)J recombination, a site-specific recombination initiated by the recombinase-activating genes 1 and 2 (RAG1/2; Matthews and Oettinger, 2009). Mature B cells bearing functional non-self reactive receptors exit into the periphery and establish the primary B cell repertoire (Rajewsky, 1996). Despite the enormous diversity generated through V(D)J recombination, B cell receptors are further diversified during immune responses, in order to establish highly specific and adapted humoral immunity (Rajewsky, 1996). This is achieved through somatic hypermutation (SHM; Di Noia and Neuberger, 2007), class switch recombination (CSR) (Chaudhuri *et al.*, 2007) and in some species (rabbits, chickens, etc) by immunoglobulin gene conversion (GCV; Arakawa and Buerstedde, 2004).

SHM modifies the affinity of the B cell receptor for the triggering antigen by introducing mainly single-base pair substitutions in the IgH and IgL variable regions (Di Noia and Neuberger, 2007). SHM requires transcription through the variable regions and occurs primarily, though not exclusively, at the RGYW sequence (R = purine, G = guanine, Y = pyrimidine, W = A or T nucleotide). GCV diversifies the rearranged variable regions by *in cis* sequence transfers templated from pseudo-genes located upstream of the rearranged variable region (Arakawa and Buerstedde, 2004). SHM and GCV are at the core of antibody affinity maturation, as they generate families of related B cell clones bearing mutated receptors, which can be selected on the basis of antigen-binding affinity (Neuberger, 2008). CSR is a DNA recombination reaction that modulates antibody effector functions by switching the antibody isotype expressed (from IgM to IgG, IgA or IgE), while preserving the antibody specificity for antigen (Chaudhuri *et al.*, 2007).

Almost 10 years ago the groups led by Tasuku Honjo and Anne Durandy demonstrated that expression of activation-induced cytidine deaminase (AID) is strictly required to initiate SHM and CSR in humans and mice (Muramatsu *et al.*, 2000; Revy *et al.*, 2000). This groundbreaking discovery provided the molecular basis for the understanding of these processes and highlighted the central role played by AID in triggering antibody diversification. AID is a fairly small enzyme of 198 amino acids bearing a cytidine deaminase catalytic domain that belongs to the APOBEC family of deaminases (Conticello *et al.*, 2007). It is composed of at least three domains. A central catalytic domain, an N-terminal domain that appears to be specifically required to initiate SHM (Shinkura *et al.*, 2004) and a C-terminal domain that is required to initiate CSR (Barreto *et al.*, 2003; Ta *et al.*, 2003). In addition, AID contains a nuclear export signal which controls its subcellular localization (which is primarily cytoplasmic; Brar *et al.*, 2004; Ito *et al.*, 2004; McBride *et al.*, 2004; Geisberger *et al.*, 2009).

At the time of its discovery, the only known protein with sequence similarity to AID was the RNA-editing enzyme APOBEC1. Hence, it was logical to propose that AID functions by deaminating cytidines in RNA (Muramatsu *et al.*, 2000). Nevertheless, the group headed by Michael Neuberger proposed a now widely accepted alternative model, in which the substrate for AID are cytosine residues in DNA rather than in RNA (Petersen-Mahrt *et al.*, 2002; Rada *et al.*, 2004). The ensuing outcome (i.e. SHM, CSR or GCV) depends on the repair pathways used to resolve the initial DNA lesion mediated by AID (Petersen-Mahrt *et al.*, 2002; Rada *et al.*, 2004).

It has now been convincingly demonstrated that AID is a single-stranded DNA (ssDNA) deaminase with little or no activity on double-stranded DNA (dsDNA; Chaudhuri *et al.*, 2007). During CSR, such ssDNA substrates for AID are probably generated by transcription through the Ig genes. Much work, a significant amount of which predates the discovery of AID, unequivocally demonstrated the essential role of transcription in CSR and SHM. The germline C_H genes (except Cδ) are organized as independent transcription units (Lutzker and Alt, 1988). In response to specific activators and cytokines, transcription is initiated from a promoter upstream of an intronic (I) exon, proceeds through switch (S) regions and terminates downstream of the C_H exons. The primary transcript is spliced to remove the intronic S region and join the I-exon to the CH exons, and is polyadenylated. The processed "sterile" or "germline" transcript does not code for any protein (Chaudhuri *et al.*, 2007).

It is now well established that transcription through S regions meets a structural requirement for AID to gain access to S regions during CSR. S regions consist of 1–12kb long-repetitive DNA elements that are characterized by a G-rich non-template strand. By virtue of the G-richness of the non-template strand, transcription through S sequences leads to the generation of higher-order DNA structures in which the transcribed RNA stably hybridizes to the template strand, looping out the non-template strand. The displaced non-template strand can remain single-stranded, in an R-loop configuration (shown to form both *in vitro* [Tian and Alt, 2000; Yu *et al.*, 2003] and *in vivo* [Yu *et al.*, 2003]), or can assume additional conformations such as stem-loops from the abundant palindromic repeats (Tashiro *et al.*, 2001), or G4 DNA (Sen and Gilbert, 1988; Dempsey *et al.*, 1999; Larson *et al.*, 2005).

That R-loops play a major role in facilitating CSR emerged from the observations that inversion of endogenous 12kb Sγ1 inhibited R-loop formation *in vitro* and was accompanied with a significant decrease in CSR to IgG1 (Shinkura *et al.*, 2003). Furthermore, replacing the Sγ1 sequence with a 1kb synthetic G-rich sequence, with no S region motifs, allowed CSR, albeit to low levels (Shinkura *et al.*, 2003). Strikingly, the ability of the synthetic sequence to mediate CSR was orientation-dependent, with CSR observed only in the orientation that favored R-loop formation (G-rich on the non-template strand; Shinkura *et al.*, 2003). Finally, *in vitro* studies showed that AID could efficiently deaminate the non-template ssDNA of transcribed S regions, providing a link between S region transcription and AID activity (Chaudhuri *et al.*, 2003; Chaudhuri *et al.*, 2004).

Despite the comprehensive view presented by the DNA deamination model, several issues regarding SHM, CSR and GCV, including specific AID targeting to Ig genes, specific targeting to variable or switch regions, substrate recognition and the handling of DNA damage remain to be fully understood. Here we review the different strategies and the progress that has been achieved in understanding the mechanisms by which AID initiates antibody diversification through the search of co-factors associating with AID.

4.2 Compartmentalization of AID

Surprisingly, although the function of AID is nuclear its subcellular localization is primarily cytoplasmic. This, along with other safeguard mechanisms has presumably evolved to minimize the pathological potential of AID activity (Casellas *et al.*, 2009). It has been clearly established that AID is efficiently exported from the nucleus through a CRM1-dependent mechanism and a nuclear export signal (NES) located in the C-terminal domain of AID (Brar *et al.*, 2004; Ito *et al.*, 2004; McBride *et al.*, 2004; Geisberger *et al.*, 2009). It has been suggested that the N-terminal domain of AID contains a canonical bi-partite nuclear localization signal (NLS; Ito *et al.*, 2004). Other groups did not observe the role of this sequence in nuclear import however, and it remained to be determined whether AID passively diffuses into the nucleus, whether it is imported by an active mechanism and whether cytoplasmic retention mechanisms enforce the cytoplasmic localization of AID. Through a series of elegant studies, Patenaude and collaborators were able to show that despite its small size AID is unable to passively diffuse into the nucleus, that nuclear translocation requires energy and that large proteins which do not diffuse into the nucleus are actively imported when fused to AID (Patenaude *et al.*, 2009). Based on a three-dimensional model of AID predicting its oligomerization it was proposed that nuclear import could be mediated by a positively-charged conformational NLS that could be recognized by the importin-α family of adaptors. Consistent with this, tagged AID was able to be precipitated by GST-tagged importin-α3, whereas import-deficient and N-terminal truncation AID mutants were not (Patenaude *et al.*, 2009). Interestingly, AID was not able to diffuse into the nucleus when importin-α-mediated nuclear import was inhibited, suggesting the existence of an active cytoplasmic retention mechanism, that appears to be dependent on the C-terminal domain of AID, but that remains to be defined at the molecular level. Finally, to test the functional relevance of nuclear import in AID function Patenaude and colleagues conducted complementation experiments by expressing wild-type and import-deficient AID mutants fused to GFP in $AID^{-/-}$ B cells and assaying for CSR by flow cytometry. In these experiments they found that impaired nuclear import results in defective CSR. Therefore, the compartmentalization of AID and its function (physiological and/or pathological) is tightly regulated by the competition between active nuclear import mediated by importin-α3 and export mediated by CRM1.

4.3 The C-Terminal Domain of AID

One of the most intriguing aspects of AID function is the C-terminal domain, which as described above, is required to export AID from the nuclear compartment. In addition this domain is exclusively required for the function of human and mouse AID in CSR but not in SHM or GCV, and has been suggested to be involved in the specific targeting of AID during CSR (Barreto *et al.*, 2003; Ta *et al.*, 2003; Shinkura *et al.*, 2004; Imai *et al.*, 2005). Hence, several laboratories have attempted to identify proteins associating with this domain to uncover its unique role in CSR.

4.3.1 *Tethering of DNA damage sensors/transducers*

CSR involves the joining of S regions. Lesions induced by AID in donor and acceptor switch regions are recognized by mismatch repair and base excision repair enzymes and processed into double-stranded DNA breaks (DSBs), which are obligate intermediates (Chaudhuri and Alt, 2004). These breaks are recognized by components of the DNA damage response including ATM (Lumsden *et al.*, 2004; Reina-San-Martin *et al.*, 2004), H2AX (Petersen *et al.*, 2001; Reina-San-Martin *et al.*, 2003),

53BP1 (Manis *et al.*, 2004; Ward *et al.*, 2004), Nbs1 (Kracker *et al.*, 2005; Reina-San-Martin *et al.*, 2005), Mre11 (Dinkelmann *et al.*, 2009) and MDC1 (Lou *et al.*, 2006), which accumulate at the IgH locus in the form of DNA repair foci (Petersen *et al.*, 2001). The recruitment of DNA damage response proteins elicited at the IgH locus during CSR has been proposed to promote changes in chromatin structure that allow for the synapsis of S regions (Reina-San-Martin *et al.*, 2003; Manis *et al.*, 2004) prior to resolution by classical and alternative non-homologous end-joining mechanisms (Casellas *et al.*, 1998; Manis *et al.*, 1998; Bosma *et al.*, 2002; Reina-San-Martin *et al.*, 2003; Soulas-Sprauel *et al.*, 2007; Yan *et al.*, 2007; Franco *et al.*, 2008; Kotnis *et al.*, 2009; Robert *et al.*, 2009). Despite robust DNA damage responses, DSBs triggered by AID can be aberrantly processed to generate oncogenic translocations involving the IgH locus and oncogenes such as c-myc (Ramiro *et al.*, 2004; Ramiro *et al.*, 2006; Robbiani *et al.*, 2008; Robbiani *et al.*, 2009; Robert *et al.*, 2009; Wang *et al.*, 2009). How precisely AID-induced DSBs are processed to favor long-range recombination while preventing genomic instability and cell transformation is a question of great interest.

In 2005 Wu and collaborators reported the interaction between AID and the catalytic subunit of DNA-PK (Wu *et al.*, 2005). This interaction was uncovered in transient transfection experiments using HeLa cells expressing an AID protein bearing an N-terminal Flag tag by affinity purification followed by SDS-PAGE and protein identification by MALDI-TOF mass spectrometry. Interactions were verified by reciprocal immunoprecipitations of the endogenous DNA-PK_{CS} and overexpressed AID. Although this association most likely takes place in the nuclear compartment as suggested (Wu *et al.*, 2005), given the stoichiometry of the association and the different subcellular localization of both proteins it is difficult to conclude whether this association is nuclear or cytoplasmic. The interaction between DNA-PK_{CS} and AID appeared to be mediated by the C-terminal domain of AID, as C-terminal truncation mutants were not able to co-precipitate DNA-PK_{CS}, whereas an N-terminal truncation mutant did. Expression of the C-terminal domain of AID (amino acids 124-198) however, did not precipitate DNA-PK_{CS} as it would have been expected. As AID catalytic mutants failed to associate with DNA-PK_{CS}, it was proposed that the association between AID and DNA-PK_{CS} occurs only after DNA damage and it was curiously proposed that DNA is a co-factor for DNA-PK_{CS} binding to AID. Expression of C-terminal truncation mutants of AID, which in these experiments do not associate with AID, resulted in persistent DNA damage and increased cell death. The model proposed (Wu *et al.*, 2005), was that AID binding to DNA through its catalytic domain results in a conformational change that exposes the C-terminal domain, which in turn recruits DNA-PK_{CS}. Following cytosine deamination, DNA-PK_{CS} dissociates from AID to assemble a NHEJ repair complex.

Using a similar strategy, we have also been able to identify DNA-PK_{CS} as one of the multiple proteins that associate with AID (Reina-San-Martin and Nussenzweig, unpublished data). In addition, we have found that other DNA damage sensors including ATM, Nbs1, Mre11 and Rad50 (but not Ku70, Ku80 or 53BP1) also associate with AID (Reina-San-Martin and Nussenzweig, unpublished data). Our results differ from those of Wu and collaborators in that all the DNA damage sensors/transducers that we have identified to be associated with AID are precipitated by a catalytic mutant and by a C-terminal truncation mutant of AID, suggesting that complex formation occurs prior to DNA damage induction independently of the C-terminal domain of AID (Reina-San-Martin and Nussenzweig, unpublished data). An alternative model is that AID associates to DNA damage sensors/transducers prior to switch region recruitment, and that this complex is tethered by AID to facilitate the detection of DNA damage, to promote long-range recombination and to reduce the risk of abnormal DNA damage resolution during CSR.

4.3.2 *MDM2*

To identify proteins directly interacting with AID and having a potential role in antibody diversification mechanisms, MacDuff and collaborators undertook a Lex-based yeast two-hybrid screen using full-length AID and a mouse spleen cDNA library. In this screen, five clones encoding MDM2 were identified as specific interactors with AID (MacDuff *et al.*, 2006). None of the clones encoded full-length MDM2 and all lacked the N-terminal half of the protein. The shortest one, which encoded amino acids 377-489 of MDM2, helped define the minimal domain of MDM2 required to interact with AID. Immunoprecipitation and western blot experiments using 293T, Ramos and DT40 cells overexpressing tagged AID confirmed the interaction, although it was only observed when Flag-tagged AID was immunoprecipitated. This suggested that the MDM2-AID interaction is only transient or weak. To define the domain in AID that is responsible for the association with MDM2 several point mutants and C-terminal truncations were assessed for interaction with MDM2. To allow for a separation of function, mutants were screened for catalytic activity in the *E. coli*-based DNA deamination assay. Of particular interest was the C-terminal truncation of AID-lacking amino acids 189-198. This mutant had increased deamination activity, and was unable to interact with MDM2 by yeast two-hybrid. A catalytic mutant of AID (E58A) was able to interact with MDM2. Based on these results it was concluded that AID associates with MDM2 independently of its catalytic activity and through its C-terminal domain. As the C-terminal domain of AID contains a robust nuclear export signal that controls its subcellular localization (Brar *et al.*, 2004; Ito *et al.*, 2004; McBride *et al.*, 2004; Geisberger *et al.*, 2009), it was proposed that MDM2 participates in the shuttling of AID between the cytoplasm and the nucleoplasm. To test this hypothesis, the chicken MDM2 gene was disrupted by homologous recombination in DT40 cells, a B cell line that constitutively diversifies its immunoglobulin genes *in vitro* by AID-dependent gene conversion (Arakawa *et al.*, 2002). Contrary to expectations, disruption of MDM2 did not result in a modification of the subcellular localization of AID. It was then proposed that MDM2 may participate in Ig gene conversion by targeting AID or by modifying its activity through ubiquitination. To explore the functional role of MDM2 in GCV the appearance of surface IgM+ clones was determined during fluctuation analysis over a 4-week period. Only a slight increase in the proportion of IgM+ cells was observed. Similarly, overexpression of MDM2 resulted in a slight reduction in surface IgM+ cells. Thus although MDM2 might negatively regulate AID activity (perhaps through its E3 ubiquitin ligase activity), it appears to be dispensable for GCV. An alternative possibility, which would be consistent with its interaction with the C-terminal domain of AID, is that MDM2 is required for CSR. However, this alternative has not been explored.

4.4 Targeting AID in the Context of Cotranscriptional Pre-mRNA Splicing by CTNNBL1

To identify proteins potentially controlling the access of AID to the nuclear compartment, Conticello and collaborators performed a cytoplasmic yeast two-hybrid screen (Conticello *et al.*, 2008). Curiously, the only clones that were consistently retrieved in this screen encoded N-terminal truncation mutants lacking an NLS of the nuclear protein CTNNBL1. Co-immunoprecipitation experiments using cytoplasmic CTNNBL1$^{\Delta NLS}$ and AID-APOBEC2 chimeric proteins fused to GFP confirmed this interaction and defined that amino acids 39–42 in AID are required for the association between both proteins. Although CTNNBL1 was identified in a cytoplasmic screen, it appears to associate with AID in the nucleus as determined by immunoprecipitation and yeast two-hybrid experiments. Amino acid replacement at positions 39–42 (AID$^{39/42}$) did not impair the catalytic activity of AID but instead resulted in a significant increase in its mutator capability, as determined in *E. coli*. Interestingly, expression of AID$^{39/42}$ in AID-deficient DT40 cells was not able to restore gene conversion at the levels observed by wild-type AID. Consistent with this, disruption of CTNNBL1 in DT40 cells resulted in a substantial reduction in the frequency of IgV diversification, although it was not completely abolished. Similarly, expression of AID$^{39/42}$ in AID-deficient B cells failed to reconstitute CSR. Failure to trigger antibody diversification by AID$^{39/42}$ was not due to impaired phosphorylation of serine 38 (Ser-38) or defective AID-RPA interaction. Thus, CTNNBL1 associates with AID and this association is required to trigger gene conversion and class switch recombination. An additional yeast two-hybrid screen using CTNNBL1 as a bait uncovered an interesting novel aspect of the CTNNLB1/AID interaction. This screen identified CDC5L as an interactor, consistent with the fact that both CTNNBL1 and CDC5L have been previously described to be present in a purified splicing complex. It was shown that AID can also be part of this complex, as it was able to associate with endogenous CDC5L. Finally, chromatin immunoprecipitation experiments using tagged CTNNBL1 indicate that this complex can associate with chromatin and suggests the exciting possibility that it might be involved in the targeting of AID to immunoglobulin loci in the context of cotranscriptional pre-mRNA splicing (Conticello *et al.*, 2008).

4.5 Replication Protein A (RPA)

While R-loop formation provides a suitable mechanism by which ssDNA substrates for AID could be generated at S regions during CSR, such a mechanism does not provide an explanation as to how AID gains access to transcribed variable regions during SHM. V genes are not G:C-rich and do not have a clear-cut propensity to form R-loops. To determine if co-factors could allow AID deamination of substrates that do not form R-loops, an *in vitro* assay was established that measured AID-mediated deamination of transcribed synthetic SHM substrates highly enriched in RGYW motifs (Chaudhuri *et al.*, 2004). It was observed that the ssDNA binding protein, RPA, interacts with AID purified from B cells via its 32kDa subunit (RPA consists of 17, 32 and 70kDa subunits) and the AID-RPA complex efficiently deaminates transcribed SHM substrates DNA (Chaudhuri *et al.*, 2004). Additional experiments showed that the AID-RPA complex, but not the two proteins separately, can bind to transcribed RGYW containing DNA *in vitro*, and strikingly, deaminations were observed at or near these hotspot motifs (Chaudhuri *et al.*, 2004). As AID preferentially deaminates RGYW sequences *in vitro* (Pham *et al.*, 2003; Yu *et al.*, 2004), and RPA can bind to ssDNA bubbles as small as 4 nucleotides (Matsunaga *et al.*, 1996), it was proposed that the AID-RPA complex binds to and stabilizes ssDNA in the context of transcription bubbles, allowing AID- mediated deamination at the RGYW sequences (Chaudhuri *et al.*, 2004).

4.6 Protein Kinase A (PKA) and Regulation of AID Activity by Phosphorylation

Mass spectrometric analysis of AID purified from splenic B cells revealed that it is phosphorylated at at least three sites, Ser-38, threonine-140 and tyrosine-184 (Basu *et al.*, 2005; McBride *et al.*, 2006; Pasqualucci *et al.*, 2006). The Ser-38 residue in AID exists in the context of a consensus cAMP-dependent protein kinase A (PKA) phosphorylation motif. AID can be phosphorylated *in vitro* by PKA at Ser-38 and AID thus phosphorylated can then bind RPA and mediate deamination of SHM substrates (Basu *et al.*, 2005; McBride *et al.*, 2006). Conversely, mutation of Ser-38 to alanine (AID^{S38A}) impaired RPA interaction and deamination of transcribed SHM substrates, without affecting deamination of ssDNA or transcribed S-regions. Thus, the interaction between AID and RPA requires phosphorylation of AID at Ser-38.

The only known biochemical consequence of AID phosphorylation at Ser-38 appears to be the activation of the RPA-binding ability of AID (Basu *et al.*, 2008). Since *in vitro* studies showed that the AID-RPA interaction is required for AID-mediated deamination of SHM substrates but not transcribed S regions, it was reasonable to predict that AID phosphorylation at Ser-38 would be required for SHM and not for CSR. However, the fact that S regions are rich in RGYW motifs and that AID-RPA deaminates non-R-loop forming *Xenopus* Sμ regions *in vitro* at sites that reflect recombination targets in mouse B cells, supported the notion that such an R-loop-independent but RGYW-dependent mechanism might be operational during CSR in mammalian cells (Zarrin *et al.*, 2004). In agreement, AID^{S38A} expressed via retroviral transduction into AID-deficient splenic B cells was only to 10–20% as active as wild-type AID in reconstituting CSR, indicating that the S38A mutation severely compromises CSR (Basu *et al.*, 2005; McBride *et al.*, 2006; Pasqualucci *et al.*, 2006).

Interpretation of retroviral transduction experiments could, however, be complicated by varying levels of protein expression, especially a hypomorph such as AID^{S38A}, that could potentially contribute to contradictory results as reported by Honjo and colleagues who failed to observe any significant effect of the S38A mutation on CSR (Shinkura *et al.*, 2007). To unequivocally address the requirement of AID phosphorylation at Ser-38 on CSR and SHM, we used gene-targeting to generate AID^{S38A} knock-in mutant mice that expressed AID^{S38A} protein from the endogenous locus (Cheng *et al.*, 2009). B cells from AID^{S38A} homozygous mutant mice failed to interact with RPA and displayed a significant defect in both CSR and SHM, consistent with the *in vitro* results (for SHM) and retroviral transduction data (for CSR). The effect of the AID^{S38A} mutation was even more pronounced in an haplo-insufficient background where AID^{S38A} was expressed from only one AID allele, with the other allele being "null" for AID (Cheng *et al.*, 2009). These results, along with very similar findings from a contemporaneous, independent study (McBride *et al.*, 2008), lay to rest the apparent controversy surrounding whether or not the AID^{S38A} mutant protein has substantially reduced CSR activity (Basu *et al.*, 2005; McBride *et al.*, 2006; Pasqualucci *et al.*, 2006; Basu *et al.*, 2007). Furthermore, several lines of evidence suggest that the AID^{S38A} phenotype is related to a defect in phosphorylating AID. First, the ssDNA deamination activity of mutant AID^{S38A} is comparable to that of wild-type AID (Basu *et al.*, 2005), indicating that the mutation does not alter the catalytic activity of the protein. Second, unphosphorylated AID does not efficiently bind RPA and mediate deamination of transcribed dsDNA *in vitro*, despite retaining ssDNA deamination activity (Basu *et al.*, 2005; Chaudhuri *et al.*, 2004). Finally, second site mutations that restore RPA binding and transcription-dependent dsDNA deamination activity to AID^{S38A} in the absence of PKA phosphorylation significantly rescue impaired CSR activity of the AID^{S38A} mutant protein (Basu *et al.*, 2008). Together, these results provide compelling evidence that impaired CSR and SHM activity of the AID^{S38A} mutant protein is due to defective AID phosphorylation and associated inability to bind RPA.

The role of RPA in CSR is not clear at this point. It is conceivable that RPA may function downstream of AID-mediated deamination, such as in the recruitment of UNG or mismatch repair proteins, which convert deaminated cytidines to DSBs (Stavnezer *et al.*, 2008). In addition, RPA might recruit proteins such as 53BP1 and H2AX to DSBs to promote synapsis between distal broken S regions prior to their ligation during CSR. The known requirements of UNG, mismatch repair proteins, 53BP1 and H2AX in CSR and the reported interactions of these proteins with RPA support a role of RPA downstream of DNA deamination (Chaudhuri *et al.*, 2007). A failure to recruit RPA to S regions could thus impair conversion of the deaminated residues into DSBs or the interaction between distal DSBs. Either or both defects could be manifested as a marked defect in CSR. Ongoing CSR observed in

AID^{S38A} probably reflects the ability of unphosphorylated AID to bind to and deaminate S regions, a few of which could be converted into nicks and DSBs and a productive recombination reaction with a downstream S region in an RPA-independent mechanism. Thus, it is possible that low frequencies of CSR could occur in the absence of assembled PKA complexes. This notion is consistent with the observation that the artificial generation of DSBs in S regions can allow CSR, albeit at reduced levels, in the absence of AID (Zarrin *et al.*, 2007).

The requirement of RPA in CSR is potentially distinct from that for SHM, where the AID–RPA interaction is required to permit AID access to transcribed V genes (Chaudhuri *et al.*, 2004). The SHM defect observed in AID^{S38A} mice supports the proposed model that RPA association with AID has some role in promoting AID access to transcribed V exons during SHM (Chaudhuri *et al.*, 2004). Whether RPA has additional roles in the processing of deaminated DNA during SHM will require detailed characterization of the SHM spectrum in the AID^{S38A} mutant B cells. That SHM does occur, albeit at reduced levels, in AID^{S38A} B cells suggests that AID must have access to V region exons in an S38 phosphorylation and RPA- independent fashion. Given that transcription is clearly an important factor in AID access to V region exons (Odegard and Schatz, 2006), one possible access mechanism, related to R-loop access, would be via spliceosome-associated factors, such as CTNNBL1 (Li and Manley, 2006; Gomez-Gonzalez and Aguilera, 2007; Conticello *et al.*, 2008). Another mechanism of AID access to V genes could be through transient changes in the DNA topology during transcription elongation by RNA polymerase II that could potentially expose ssDNA structures (Shen and Storb, 2004).

4.7 Recruitment of PKA to Switch Region Sequences

The mechanism that directs AID-induced DSBs to a very restricted region of the genome, namely the S regions during CSR, remains enigmatic. Our recent results on the recruitment of PKA to S regions in activated B cells could potentially shed light on the mechanism by which AID activity is specified at the Ig locus. Chromatin immunoprecipitation experiments in primary splenic B cells showed that AID can bind to S regions independent of its phosphorylation status at Ser-38 (Vuong *et al.*, 2009). Thus, AID^{S38A} is as competent to bind S regions as wild-type AID. We also find that in activated B cells, both the regulatory RIα and catalytic Cα subunits of PKA are specifically associated with S regions activated to undergo CSR (Vuong *et al.*, 2009). The binding of PKA subunits to S regions is independent of AID. These observations prompted us to propose that CSR requires the assembly of a complex consisting of the PKA catalytic and regulatory subunits and AID on S regions. A burst of cyclic-AMP releases activates the catalytic subunit of PKA which phosphorylates AID at Ser-38, thereby recruiting RPA and initiating the CSR cascade at these sites (Vuong *et al.*, 2009). Consistent with these findings, conditional inactivation of PKA disrupts CSR due to markedly reduced AID phosphorylation and RPA recruitment (Vuong *et al.*, 2009). Thus, according to this model, PKA nucleates the formation of active AID complexes on S regions, leading to the generation of a high density of DNA lesions required for CSR.

The PKA-dependent activation of AID has several potential implications in preventing inadvertent DNA lesions. First, since AID is in a complex with both the catalytic and regulatory PKA subunits (Vuong *et al.*, 2009), it is possible that the sequestration of AID in an inactive PKA holoenzyme complex could effectively limit the total cellular concentration of active AID. Second, the requirement for the presence of both PKA and AID at S regions suggests that DSB-promoting activity of AID is unleashed only within the context of *bona fide* targets. Thus, while AID can deaminate and mutate several transcribed genes in germinal center B cells at a low rate (Liu *et al.*, 2008), its full DSB-generating activity as required for CSR is unmasked only when it is co-recruited with PKA, thereby effectively restricting AID-dependent mutagenesis at non-Ig genes. Finally, the AID-independent recruitment of PKA to activated S regions (Vuong *et al.*, 2009) and the ability of AID to interact with PKA (Basu *et al.*, 2005; Pasqualucci *et al.*, 2006; Vuong *et al.*, 2009) lead us to consider the possibility that PKA itself could be the factor that targets AID to S regions during CSR.

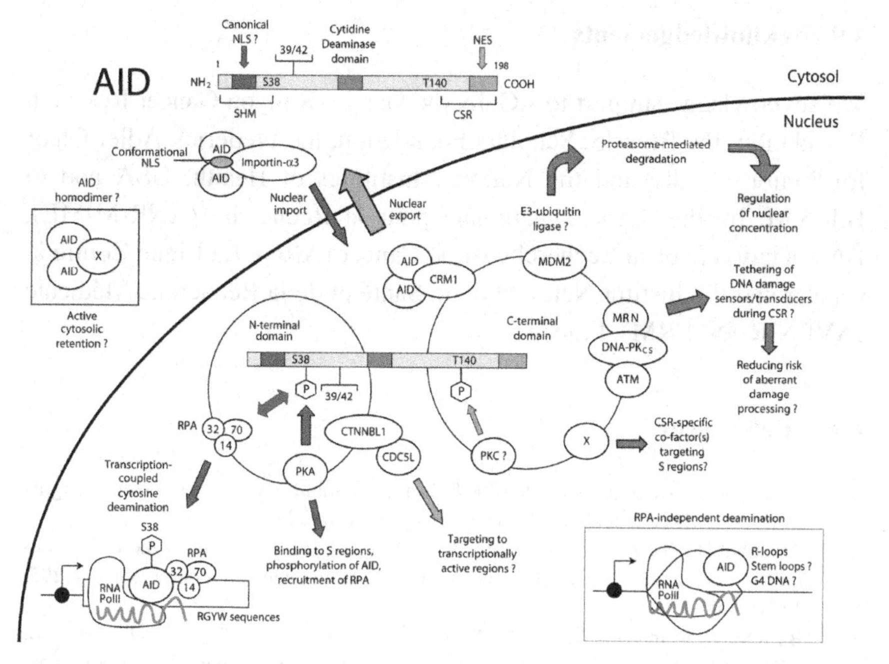

Figure 4.1. **Partners in diversity.** Schematic representation of the role of the different co-factors identified for AID in regulating AID function.

4.8 Concluding Remarks

The major challenges faced in the search for AID co-factors and the study of their specific roles in diversifying the B cell receptor repertoire, mainly arise from the relatively low levels of endogenous AID expression and the particular propensity of AID to non-specifically associate with multiple proteins both in biochemical assays and in yeast two-hybrid experiments. Despite these difficulties, we have been able to gain substantial insight into the multiple levels of AID regulation and the different mechanisms that target AID to immunoglobulin loci (summarized in Fig. 4.1). It is becoming increasingly apparent that the benefit acquired by receptor diversification requires the establishment of tight regulatory mechanisms, which need to be carefully maintained to limit the pathological potential of AID activity.

4.9 Acknowledgements

We acknowledge support to J.C. by the Damon Runyon Cancer Research Foundation, the Bressler Scholars Foundation, the Frederick Adler Chair for Junior Faculty and the National Institutes of Health, USA and to B.R.S.M. by the Agence Nationale pour la Recherche (ANR-MIME), l'Association pour la Recherche sur le Cancer (ARC), La Ligue Contre le Cancer and the Institut National de la Santé et de la Recherche Médicale (AVENIR-INSERM), France.

4.10 References

1. Arakawa H., Buerstedde J.M. (2004). Immunoglobulin gene conversion: insights from bursal B cells and the DT40 cell line. *Dev Dyn* **229**: 458–464.
2. Arakawa H., Hauschild J., Buerstedde J.M. (2002). Requirement of the activation-induced deaminase (AID) gene for immunoglobulin gene conversion. *Science* **295**: 1301–1306.
3. Barreto V., Reina-San-Martin B., Ramiro A.R. *et al.* (2003). C-terminal deletion of AID uncouples class switch recombination from somatic hypermutation and gene conversion. *Mol Cell* **12**: 501–508.
4. Basu U., Chaudhuri J., Alpert C. *et al.* (2005). The AID antibody diversification enzyme is regulated by protein kinase A phosphorylation. *Nature* **438**: 508–511.
5. Basu U., Chaudhuri J., Phan R.T. *et al.* (2007). Regulation of activation induced deaminase via phosphorylation. *Adv Exp Med Biol* **596**: 129–137.
6. Basu U., Wang Y., Alt F.W. (2008). Evolution of phosphorylation-dependent regulation of activation-induced cytidine deaminase. *Mol Cell* **32**: 285–291.
7. Bosma G.C., Kim J., Urich T. *et al.* (2002). DNA-dependent protein kinase activity is not required for immunoglobulin class switching. *J Exp Med* **196**: 1483–1495.
8. Brar S.S., Watson M., Diaz M. (2004). Activation-induced cytosine deaminase (AID) is actively exported out of the nucleus but retained by the induction of DNA breaks. *J Biol Chem* **279**: 26395–26401.
9. Casellas R., Nussenzweig A., Wuerffel R. *et al.* (1998). Ku80 is required for immunoglobulin isotype switching. *EMBO J* **17**: 2404–2411.
10. Casellas R., Yamane A., Kovalchuk A.L. *et al.* (2009). Restricting activation-induced cytidine deaminase tumorigenic activity in B lymphocytes. *Immunology* **126**: 316–328.
11. Chaudhuri J., Alt F.W. (2004). Class-switch recombination: interplay of transcription, DNA deamination and DNA repair. *Nat Rev Immunol* **4**: 541–552.
12. Chaudhuri J., Basu U., Zarrin A. *et al.* (2007). Evolution of the immunoglobulin heavy chain class switch recombination mechanism. *Adv Immunol* **94**: 157–214.

13. Chaudhuri J., Khuong C., Alt F.W. (2004). Replication protein A interacts with AID to promote deamination of somatic hypermutation targets. *Nature* **430**: 992–998.
14. Chaudhuri J., Tian M., Khuong C. *et al.* (2003). Transcription-targeted DNA deamination by the AID antibody diversification enzyme. *Nature* **422**: 726–730.
15. Cheng H.L., Vuong B.Q., Basu U. *et al.* (2009). Integrity of the AID serine-38 phosphorylation site is critical for class switch recombination and somatic hypermutation in mice. *Proc Natl Acad Sci USA* **106**: 2717–2722.
16. Conticello S.G., Ganesh K., Xue K. *et al.* (2008). Interaction between antibody-diversification enzyme AID and spliceosome-associated factor CTNNBL1. *Mol Cell* **31**: 474–484.
17. Conticello S.G., Langlois M.A., Yang Z. *et al.* (2007). DNA deamination in immunity: AID in the context of its APOBEC relatives. *Adv Immunol* **94**: 37–73.
18. Dempsey L.A., Sun H., Hanakahi L.A. *et al.* (1999). G4 DNA binding by LR1 and its subunits, nucleolin and hnRNP D, a role for G-G pairing in immunoglobulin switch recombination. *J Biol Chem* **274**: 1066–1071.
19. Di Noia J.M., Neuberger M.S. (2007). Molecular mechanisms of antibody somatic hypermutation. *Annu Rev Biochem* **76**: 1–22.
20. Dinkelmann M., Spehalski E., Stoneham T. *et al.* (2009). Multiple functions of MRN in end-joining pathways during isotype class switching. *Nat Struct Mol Biol* **16**: 808–813.
21. Franco S., Murphy M.M., Li G. *et al.* (2008). DNA-PKcs and Artemis function in the end-joining phase of immunoglobulin heavy chain class switch recombination. *J Exp Med* **205**: 557–564.
22. Geisberger R., Rada C., Neuberger M.S. (2009). The stability of AID and its function in class-switching are critically sensitive to the identity of its nuclear-export sequence. *Proc Natl Acad Sci USA* **106**: 6736–6741.
23. Gomez-Gonzalez B., Aguilera A. (2007). Activation-induced cytidine deaminase action is strongly stimulated by mutations of the THO complex. *Proc Natl Acad Sci USA* **104**: 8409–8414.
24. Imai K., Zhu Y., Revy P. *et al.* (2005). Analysis of class switch recombination and somatic hypermutation in patients affected with autosomal dominant hyper-IgM syndrome type 2. *Clin Immunol* **115**: 277–285.
25. Ito S., Nagaoka H., Shinkura R. *et al.* (2004). Activation-induced cytidine deaminase shuttles between nucleus and cytoplasm like apolipoprotein B mRNA editing catalytic polypeptide 1. *Proc Natl Acad Sci USA* **101**: 1975–1980.
26. Kotnis A., Du L., Liu C. *et al.* (2009). Non-homologous end joining in class switch recombination: the beginning of the end. *Philos Trans R Soc Lond B Biol Sci* **364**: 653–665.
27. Kracker S., Bergmann Y., Demuth I. *et al.* (2005). Nibrin functions in Ig class-switch recombination. *Proc Natl Acad Sci USA* **102**: 1584–1589.
28. Larson E.D., Duquette M.L., Cummings W.J. *et al.* (2005). MutSalpha binds to and promotes synapsis of transcriptionally activated immunoglobulin switch regions. *Curr Biol* **15**: 470–474.

29. Li X., Manley J.L. (2006). Cotranscriptional processes and their influence on genome stability. *Genes Dev* **20**: 1838–1847.
30. Liu M., Duke J.L., Richter D.J. *et al.* (2008). Two levels of protection for the B cell genome during somatic hypermutation. *Nature* **451**: 841–845.
31. Lou Z., Minter-Dykhouse K., Franco S. *et al.* (2006). MDC1 maintains genomic stability by participating in the amplification of ATM-dependent DNA damage signals. *Mol Cell* **21**: 187–200.
32. Lumsden J.M., McCarty T., Petiniot L.K. *et al.* (2004). Immunoglobulin class switch recombination is impaired in Atm-deficient mice. *J Exp Med* **200**: 1111–1121.
33. Lutzker S., Alt F.W. (1988). Structure and expression of germ line immunoglobulin gamma 2b transcripts. *Mol Cell Biol* **8**: 1849–1852.
34. MacDuff D.A., Neuberger M.S., Harris R.S. (2006). MDM2 can interact with the C-terminus of AID but it is inessential for antibody diversification in DT40 B cells. *Mol Immunol* **43**: 1099–1108.
35. Manis J.P., Gu Y., Lansford R. *et al.* (1998). Ku70 is required for late B cell development and immunoglobulin heavy chain class switching. *J Exp Med* **187**: 2081–2089.
36. Manis J.P., Morales J.C., Xia Z. *et al.* (2004). 53BP1 links DNA damage-response pathways to immunoglobulin heavy chain class-switch recombination. *Nat Immunol* **5**: 481–487.
37. Matsunaga T., Park C.H., Bessho T. *et al.* (1996). Replication protein A confers structure-specific endonuclease activities to the XPF-ERCC1 and XPG subunits of human DNA repair excision nuclease. *J Biol Chem* **271**: 11047–11050.
38. Matthews A.G., Oettinger M.A. (2009). RAG: a recombinase diversified. *Nat Immunol* **10**: 817–821.
39. McBride K.M., Barreto V., Ramiro A.R. *et al.* (2004). Somatic hypermutation is limited by CRM1-dependent nuclear export of activation-induced deaminase. *J Exp Med* **199**: 1235–1244.
40. McBride K.M., Gazumyan A., Woo E.M. *et al.* (2006). Regulation of hypermutation by activation-induced cytidine deaminase phosphorylation. *Proc Natl Acad Sci USA* **103**: 8798–8803.
41. McBride K.M., Gazumyan A., Woo E.M. *et al.* (2008). Regulation of class switch recombination and somatic mutation by AID phosphorylation. *J Exp Med* **205**: 2585–2594.
42. Muramatsu M., Kinoshita K., Fagarasan S. *et al.* (2000). Class switch recombination and hypermutation require activation-induced cytidine deaminase (AID), a potential RNA editing enzyme. *Cell* **102**: 553–563.
43. Neuberger M.S. (2008). Antibody diversification by somatic mutation: from Burnet onwards. *Immunol Cell Biol* **86**: 124–132.
44. Odegard V.H., Schatz D.G. (2006). Targeting of somatic hypermutation. *Nat Rev Immunol* **6**: 573–583.
45. Pasqualucci L., Kitaura Y., Gu H. *et al.* (2006). PKA-mediated phosphorylation regulates the function of activation-induced deaminase (AID) in B cells. *Proc Natl Acad Sci USA* **103**: 395–400.

46. Patenaude A.M., Orthwein A., Hu Y. *et al.* (2009). Active nuclear import and cytoplasmic retention of activation-induced deaminase. *Nat Struct Mol Biol* **16**: 517–527.
47. Petersen-Mahrt S.K., Harris R.S., Neuberger M.S. (2002). AID mutates E. coli suggesting a DNA deamination mechanism for antibody diversification. *Nature* **418**: 99–103.
48. Petersen S., Casellas R., Reina-San-Martin B. *et al.* (2001). AID is required to initiate Nbs1/gamma-H2AX focus formation and mutations at sites of class switching. *Nature* **414**: 660–665.
49. Pham P., Bransteitter R., Petruska J. *et al.* (2003). Processive AID-catalysed cytosine deamination on single-stranded DNA simulates somatic hypermutation. *Nature* **424**: 103–107.
50. Rada C., Di Noia J.M., Neuberger M.S. (2004). Mismatch recognition and uracil excision provide complementary paths to both Ig switching and the A/T-focused phase of somatic mutation. *Mol Cell* **16**: 163–171.
51. Rajewsky K. (1996). Clonal selection and learning in the antibody system. *Nature* **381**: 751–758.
52. Ramiro A.R., Jankovic M., Callen E. *et al.* (2006). Role of genomic instability and p53 in AID-induced c-myc-Igh translocations. *Nature* **440**: 105–109.
53. Ramiro A.R., Jankovic M., Eisenreich T. *et al.* (2004). AID is required for c-myc/IgH chromosome translocations in vivo. *Cell* **118**: 431–438.
54. Reina-San-Martin B., Chen H.T., Nussenzweig A. *et al.* (2004). ATM is required for efficient recombination between immunoglobulin switch regions. *J Exp Med* **200**: 1103–1110.
55. Reina-San-Martin B., Difilippantonio S., Hanitsch L. *et al.* (2003). H2AX is required for recombination between immunoglobulin switch regions but not for intra-switch region recombination or somatic hypermutation. *J Exp Med* **197**: 1767–1778.
56. Reina-San-Martin B., Nussenzweig M.C., Nussenzweig A. *et al.* (2005). Genomic instability, endoreduplication, and diminished Ig class-switch recombination in B cells lacking Nbs1. *Proc Natl Acad Sci USA* **102**: 1590–1595.
57. Revy P., Muto T., Levy Y. *et al.* (2000). Activation-induced cytidine deaminase (AID) deficiency causes the autosomal recessive form of the Hyper-IgM syndrome (HIGM2). *Cell* **102**: 565–575.
58. Robbiani D.F., Bothmer A., Callen E. *et al.* (2008). AID is required for the chromosomal breaks in c-myc that lead to c-myc/IgH translocations. *Cell* **135**: 1028–1038.
59. Robbiani D.F., Bunting S., Feldhahn N. *et al.* (2009). AID produces DNA double-strand breaks in non-Ig genes and mature B cell lymphomas with reciprocal chromosome translocations. *Mol Cell* **36**: 631–641.
60. Robert I., Dantzer F., Reina-San-Martin B. (2009). Parp1 facilitates alternative NHEJ, whereas Parp2 suppresses IgH/c-myc translocations during immunoglobulin class switch recombination. *J Exp Med* **206**: 1047–1056.
61. Sen D., Gilbert W. (1988). Formation of parallel four-stranded complexes by guanine-rich motifs in DNA and its implications for meiosis. *Nature* **334**: 364–366.

62. Shen H.M., Storb U. (2004). Activation-induced cytidine deaminase (AID) can target both DNA strands when the DNA is supercoiled. *Proc Natl Acad Sci USA* **101**: 12997–13002.
63. Shinkura R., Ito S., Begum N.A. *et al.* (2004). Separate domains of AID are required for somatic hypermutation and class-switch recombination. *Nat Immunol* **5**: 707–712.
64. Shinkura R., Okazaki I.M., Muto T. *et al.* (2007). Regulation of AID function in vivo. *Adv Exp Med Biol* **596**: 71–81.
65. Shinkura R., Tian M., Smith M. *et al.* (2003). The influence of transcriptional orientation on endogenous switch region function. *Nat Immunol* **4**: 435–441.
66. Soulas-Sprauel P., Le Guyader G., Rivera-Munoz P. *et al.* (2007). Role for DNA repair factor XRCC4 in immunoglobulin class switch recombination. *J Exp Med* **204**: 1717–1727.
67. Stavnezer J., Guikema J.E., Schrader C.E. (2008). Mechanism and regulation of class switch recombination. *Annu Rev Immunol* **26**: 261–292.
68. Ta V.T., Nagaoka H., Catalan N. *et al.* (2003). AID mutant analyses indicate requirement for class-switch-specific cofactors. *Nat Immunol* **4**: 843–848.
69. Tashiro J., Kinoshita K., Honjo T. (2001). Palindromic but not G-rich sequences are targets of class switch recombination. *Int Immunol* **13**: 495–505.
70. Tian M., Alt F.W. (2000). Transcription-induced cleavage of immunoglobulin switch regions by nucleotide excision repair nucleases in vitro. *J Biol Chem* **275**: 24163–24172.
71. Vuong B.Q., Lee M., Kabir S. *et al.* (2009). Specific recruitment of protein kinase A to the immunoglobulin locus regulates class-switch recombination. *Nat Immunol* **10**: 420–426.
72. Wang J.H., Gostissa M., Yan C.T. *et al.* (2009). Mechanisms promoting translocations in editing and switching peripheral B cells. *Nature* **460**: 231–236.
73. Ward I.M., Reina-San-Martin B., Olaru A. *et al.* (2004). 53BP1 is required for class switch recombination. *J Cell Biol* **165**: 459–464.
74. Wu X., Geraldes P., Platt J.L. *et al.* (2005). The double-edged sword of activation-induced cytidine deaminase. *J Immunol* **174**: 934–941.
75. Yan C.T., Boboila C., Souza E.K. *et al.* (2007). IgH class switching and translocations use a robust non-classical end-joining pathway. *Nature* **449**: 478–482.
76. Yu K., Chedin F., Hsieh C.L. *et al.* (2003). R-loops at immunoglobulin class switch regions in the chromosomes of stimulated B cells. *Nat Immunol* **4**: 442–451.
77. Yu K., Huang F.T., Lieber M.R. (2004). DNA substrate length and surrounding sequence affect the activation-induced deaminase activity at cytidine. *J Biol Chem* **279**: 6496–6500.
78. Zarrin A.A., Alt F.W., Chaudhuri J. *et al.* (2004). An evolutionarily conserved target motif for immunoglobulin class-switch recombination. *Nat Immunol* **5**: 1275–1281.
79. Zarrin A.A., Del Vecchio C., Tseng E. *et al.* (2007). Antibody class switching mediated by yeast endonuclease-generated DNA breaks. *Science* **315**: 377–381.

Chapter 5

Resolution of AID Lesions in Class Switch Recombination

Kefei Yu[1] *and Michael R. Lieber*[2]

[1]*Microbiology and Molecular Genetics, Michigan State University*
East Lansing, MI 48824, USA
E-mail: yuke@msu.edu

[2]*USC Norris Comprehensive Cancer Ctr.*
Los Angeles, CA 90089-9176, USA
E-mail: lieber@usc.edu

AID initiates the process by which immunoglobulin heavy chain loci are recombined to permit the isotype switch from IgM to IgG, IgA and IgE, called class switch recombination. Here we discuss the current state of knowledge of AID action at class switch regions from the initial AID deamination sites to the rejoining of the DNA ends. The relative roles of base excision repair versus mismatch repair are discussed. The contribution of R-loops not only to the initial AID targeting, but also to the access of AID to the template strand, are examined. Finally, we consider the relative contribution of non-homologous end joining (NHEJ) versus alternative NHEJ enzymes for the rejoining of the double-strand DNA breaks.

5.1 Introduction

One can consider the topic of this chapter in two phases. In the first phase, AID lesions must be converted into double-strand DNA breaks. In the second phase, the double-strand breaks must be repaired for class switch recombination (CSR) to be completed.

5.2 Conversion of AID Lesions to Double-Strand DNA Breaks

5.2.1 *Uracils in switch region DNA*

Biochemical studies using purified recombinant AID proteins demonstrated unequivocally that AID is a single-stranded deoxycytidine deaminase, which is the critical activity in both CSR and in somatic hypermutation (SHM; Bransteitter *et al.*, 2003; Pham *et al.*, 2003; Dickerson *et al.*, 2003; Yu *et al.*, 2004). However, evidence for the existence of uracils specifically in class switch region DNA *in vivo* is still indirect. Much of our knowledge of AID action in class switch recombination (CSR) came from genetic dissection of pathways involved in repair of uracils in DNA (Petersen-Mahrt *et al.*, 2002; Di Noia and Neuberger, 2002; Di Noia *et al.*, 2007). Uracil is a component of the RNA, not DNA, but could be misincorporated into the DNA or converted from deoxycytidine via spontaneous deamination. Uracils are mainly repaired by the base excision repair (BER) pathway, which is initiated upon uracil removal by a uracil glycosylase (e.g. UNG2) followed by apurinic/apyrimidinic endonuclease (APE1)-mediated strand incision at the abasic site. AID-mediated cytidine deamination creates a U:G mismatch that can also be repaired by the mismatch repair (MMR) pathway, which include factors such as MSH2, MSH6, MLH1, PMS2 and Exo1 (Ehrenstein and Neuberger, 1999; Neuberger *et al.*, 2005). Consistent with the idea that uracils are critical intermediates that cause DNA lesions at class switch regions, defects of UNG2 or any of the listed MMR factors inhibits CSR to varying degrees. Importantly, knockout of both UNG2 (BER) and MSH2 (MMR) in the mouse completely abolishes CSR and forces DNA replication over uracils to give rise exclusively to C to T mutations within and downstream of VJ and VDJ (for SHM) and at class switch regions (for CSR; Rada *et al.*, 2004). Thus, the $UNG2^{-/-}MSH2^{-/-}$ mouse has been exploited to study AID footprints in the switch regions. PCR products derived from either the template or non-template DNA strands show evidence of such AID footprints (C to T mutations; Xue *et al.*, 2006). However, due to the separation of the two strands in PCR, this analysis does not allow the assembly of AID footprints on both strands of the same molecule. Such information would be very valuable for unraveling the mystery of how uracil repair leads to DNA breaks in the switch region.

5.2.2 *Base excision repair in class switch recombination*

Uracil glycosylase-initiated BER is the major pathway for repair of uracils incorporated into DNA. Although there are several glycosylases in mammalian cells that can excise uracil bases (some in specific settings), only UNG2 seems to be directly involved in CSR. Although mice deficient in UNG2 have substantial serum levels of switched immunoglobulin isotypes, CSR *in vitro* in isolated spleen B cells is severely inhibited (Rada *et al.*, 2002). Human patients harboring UNG2 mutations seem to have an even more severe CSR defect (Imai *et al.*, 2003). In addition, expression of a specific UNG2 inhibitor (ugi) in a B cell line inhibits CSR *in vitro* (Begum *et al.*, 2004). There has been some debate about whether uracil excision is directly involved in CSR as certain catalytic site mutants of UNG2 can support CSR when introduced into UNG2$^{-/-}$ B cells by retroviral transduction (Begum *et al.*, 2004). The caveat is that these mutants still have residual catalytic activity, which could be significant when expressed 100-fold over the endogenous level as a result of retroviral transduction (Di Noia and Neuberger, 2007).

In contrast to UNG2, the role of APE1 in CSR is less certain. APE1 is estimated to account for >95%, if not all, endonuclease activity at abasic sites *in vivo* (Demple and Harrison, 1994). It seems obvious that APE1 would be responsible for strand incision required for CSR, given that the role of UNG2 is established. However, because APE1 is essential for mouse embryo development and even the viability of somatic cells, it has been difficult to directly test APE1 in CSR by the conventional reverse genetic approaches. Mouse spleen B cells that are haplodeficient for APE1 have a small reduction of CSR (Guikema *et al.*, 2007), but depletion of APE1 in a cultured B cell line by siRNA technology showed no effect on CSR (Sabouri *et al.*, 2009). Confirmation of the nuclease in CSR must await additional novel experimental strategies.

5.2.3 *Mismatch repair in class switch recombination*

Many MMR components (MSH2, MSH6, MLH1, PMS2, Exo1) are found to be required for fully efficient CSR (Ehrenstein and Neuberger, 1999; Vora *et al.*, 1999; Schrader *et al.*, 1999; Bardwell *et al.*, 2004; Peron *et al.*, 2008; Martin *et al.*, 2003), suggesting that the AID-generated U:G mismatch is also processed, in part, by MMR. It is worth pointing out that MSH6, but not MSH3, was found to affect CSR efficiency (Li *et al.*, 2004). This is consistent with the finding that the MSH2/6 complex detects single-base mismatches while the MSH2/3 complex detects larger mismatches. Because of the robust activity of UNG2, MMR is generally considered as a minor (or backup) pathway for repair of uracils in DNA. It is therefore not surprising that MMR deficiency causes a milder CSR defect (two- to five-fold).

The important question is then, why does CSR require MMR at all, assuming that AID-generated uracils do not overwhelm the UNG2-dependent pathway. The key might be the different mode of action in processing of U:G mismatch. BER is a single-base repair and MMR is a patch repair pathway. In SHM, generation of a patch is critical for an error-prone re-synthesis that could spread the mutations from the initial AID hit. In CSR, the importance of such a patch is less clear, but a model has been proposed whereby generating a patch (by Exo1) might facilitate the creation of DNA double-strand breaks from scattered single-stranded nicks (Min *et al.*, 2003; Stavnezer and Schrader, 2006). Though Exo1 can likely function at some level alone, its efficiency of loading onto a DNA substrate may be significantly improved by the other MMR components.

5.2.4 *Generation of DNA double-strand breaks in switch regions*

A large gap in our knowledge of CSR is how uracil repair leads to DNA double-strand breaks in the switch region (a challenging issue, especially since we do not know the relative distribution of uracils for both DNA strands of the same duplex). The key to this question might lie in the sequence motifs in the switch regions. Transcription through switch regions tends to induce DNA secondary structures known as R-loops that expose single-stranded regions on the non-template DNA strand (Yu *et al.*, 2003). These single-stranded regions might play a critical role in facilitating AID-mediated deamination as AID is a single-strand-specific enzyme. By this model, it would be expected that the displaced non-template DNA strand has a larger load of AID-deamination events. However, data from $UNG^{-/-}MSH2^{-/-}$ mice showed that the template strand is also deaminated by AID (Xue *et al.*, 2006).

There are many possible mechanisms by which the template strand could become single-stranded and therefore become a suitable substrate for AID. One is by the complete or partial removal of the RNA transcript by cellular RNase H activities after R-loop formation (Fig. 5.1). It is conceivable that some DNA breaks could result by coincidence, from closely positioned nicks across the two DNA strands. Switch regions are known to be important for CSR, but not absolutely required (Luby *et al.*, 2001). It remains to be determined what distinguishes a switch region from any other sequence in the genome, as only switch regions support maximal CSR efficiency (Zarrin *et al.*, 2004). One noticeable feature about a switch region is that it is enriched with AGCT sequence motif (Zarrin *et al.*, 2004). AGCT is one of four WGCW (W = A or T) sequences that represent AID-preferred sites (WRC) on both DNA strands. One possible scenario is that AID, if it functions as a dimer, may introduce two nicks directly across from one another on the two DNA strands, resulting in a double-strand break (Yu *et al.*, 2004; Min *et al.*, 2005). AID, UNG2 and APE1 would be sufficient to generate a double-strand break by this mechanism.

The question then remains, why would CSR require MMR for full efficiency? Recent studies on Sμ deletion in mice have shed some light on this question. Deletion of Sμ tandem repeats (designated ΔSμTR) only mildly reduce CSR efficiency (11–63%, depending on the acceptor switch region; Luby *et al.*, 2001), suggesting that intronic sequence outside the Sμ tandem repeats have considerable CSR activity. It is rather interesting that the remaining CSR activity in ΔSμTR mice is completely dependent on the MMR (Min *et al.*, 2003). This has led to a hypothesis that in the absence of AGCT sites, AID-inflicted nicks are

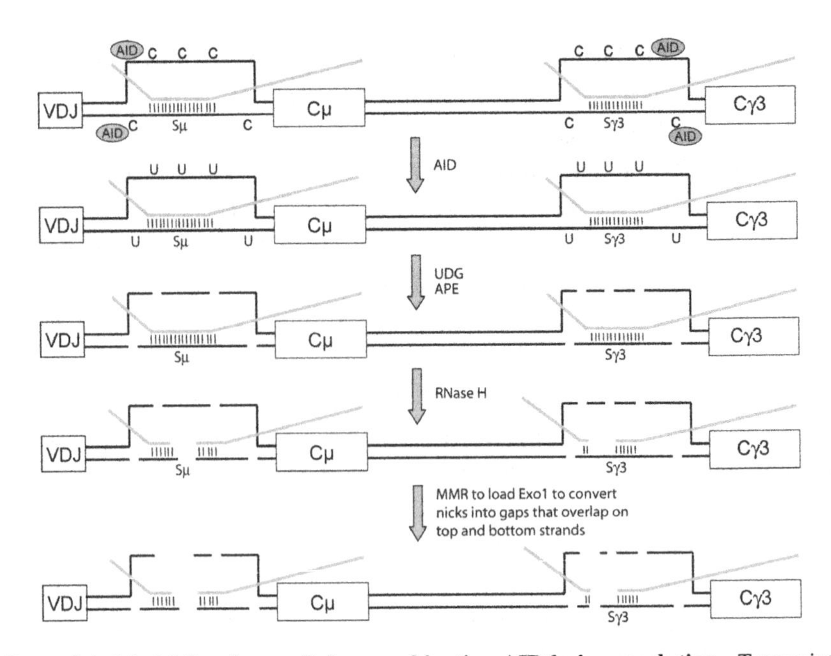

Figure 5.1. **Model for class switch recombination AID lesion resolution.** Transcripts initiating from the I exon generate R-loops located within and near the repetitive switch region. This provides single-strandedness on the non-template strand, at which AID can now act. AID converts C to U. UNG removes the base of uracils. APE nicks 5′ to abasic sites. RNase H endonucleolytically removes portions of the RNA of the RNA:DNA hybrid of the R-loop. This exposes regions of single-strandedness on the template DNA strand, thereby allowing AID to act on the template strand as well. (AID can also act on the template strand by two other means. First, AID can act at the edges of the R-loop. Second, when RNase H removes RNA spanning several switch repeats, the two DNA strands may anneal in a mis-aligned configuration, resulting in single-stranded heterologous loops on the two DNA strands; AID can act on these loops as well.) The nicks produced by AID/UNG/APE on the two DNA strands may be close together, especially within the repetitive switch region. But even when they are proximal, they may still be far enough apart (10bp) so as not to yield a double-strand break. When the AID/UNG/APE lesions are outside of the repetitive switch region, the lesions on the non-template and template strands may be even further apart (hundreds of base pairs). In that context, Exo1 may be important to resect 5′ to 3′ on the two DNA strands so that the lesions on these strands can yield a double-strand break. The MMR proteins may be important for improving the otherwise limited efficiency of Exo1 in targeting these regions. After the double-strand breaks are formed, NHEJ can then join the ends. In a minority of cases or in experimental mutants, other enzymes may participate in the joining (alternative NHEJ).

scarce and far apart, and MMR is required to connect these distal nicks via Exo1-mediated strand degradation to create a double-strand break (Min *et al.*, 2003; Stavnezer and Schrader, 2006; Eccleston *et al.*, 2009). Regardless of how the data are interpreted, the post-AID events remain an area of vigorous experimental testing and continued study.

5.3 Repair of Double-Strand DNA Breaks in Class Switch Recombination

The repair of double-strand DNA breaks generated by class switch recombination can occur by non-homologous end joining (NHEJ) or by variants of NHEJ in which some components are substituted by others or the joining occurs without certain NHEJ components.

5.3.1 *Ku and the initial phase of NHEJ*

NHEJ is sometimes called classical NHEJ. We prefer to simply use the designation NHEJ. NHEJ is thought to begin with the binding of Ku to the two DNA ends at a double-strand break, based on its high affinity for DNA ends (K_D~10^{-9}M) and its nuclear abundance, which is thought to be ~400,000 molecules. Ku is a heterodimer, consisting of Ku70 and Ku86 (often called Ku80); Ku70 has a molecular weight (MW) of 70kDa, and Ku86 has a MW of 83kDa.

Once bound to a DNA end, Ku improves the binding affinity of all of the other NHEJ components. In this sense, Ku serves as a tool belt protein, like PCNA, in that it can load the nuclease, polymerases or ligase for NHEJ in any order and repeatedly. Though Ku can improve the affinity of each of the enzymatic components for NHEJ, the other enzymatic components each have their own ability to bind and act at a DNA end without Ku. In this regard, Ku is a key facilitator, but is not essential for NHEJ. Notably, it has been shown that mice deficient for Ku have reduced efficiency of CSR (Casellas *et al.*, 1998; Manis *et al.*, 1998).

5.3.2 *Nucleases for NHEJ*

Artemis:DNA-PKcs has endonuclease activity that includes action at both 5′ and 3′ overhangs. Artemis and DNA-PKcs exist as a complex within mammalian cells. Though DNA-PKcs can bind to DNA ends by itself (or with Artemis) with a $K_D \sim 3 \times 10^{-9}$ M, the binding of DNA-PKcs to Ku:DNA end sites has a $K_D \sim 3 \times 10^{-11}$ M. Therefore, the affinity for loading of Artemis:DNA-PKcs to DNA ends is increased about 100-fold when Ku is bound to the DNA end.

Using Artemis knock-out mice, it has been shown that Artemis functions in NHEJ of CSR double-strand breaks (Franco *et al.*, 2008). In addition, the Alt laboratory has shown that DNA-PKcs is important for NHEJ of CSR double-strand breaks (Rooney *et al.*, 2005). The Bosma laboratory did not see a CSR phenotype in DNA-PKcs mutant mice (Bosma *et al.*, 2002). Methodological differences for assessing the level of CSR may account for this difference.

5.3.3 *Polymerases for NHEJ*

Polymerase μ (pol μ) and polymerase λ (pol λ) are able to function in NHEJ, based on work with purified systems and with crude extracts. These two polymerases are both members of the Pol X family of polymerases, along with polymerase β and TdT. All members of the Pol X family except polymerase β have N-terminal BRCT domains. Ku is able to recruit pol μ and pol λ in a manner that is dependent on the BRCT domain.

In vivo, absence of pol μ or pol λ does have an impact on V(D)J recombination junctions. However, there have not yet been cellular or animal studies on the effect on CSR when pol μ or pol λ are missing. In yeast, lack of POL4, the *S. cerevisiae* Pol X polymerase alters the nature of the resulting NHEJ events, but NHEJ still occurs, presumably due to some of the replicative or repair polymerases. Moreover, given the mild phenotypic effect on V(D)J recombination, lack of pol μ and/or pol λ may have relatively small effects on CSR, and are likely not to be detectable.

5.3.4 *Ligases for NHEJ*

A complex of XLF (also called Cernunnos), XRCC4 and DNA ligase IV is the primary ligase activity for NHEJ. In yeast, in the absence of DNA ligase IV, the rejoining of DNA ends is at least 10-fold less efficient, and the DNA ends that are joined appear to rely on longer than usual regions of terminal microhomology (greater than four base pairs; Daley *et al.*, 2005). In mammalian cells, survival of cells after treatment with ionizing radiation appears to be reduced about 10-fold when they lack ligase IV (Riballo *et al.*, 2004).

Mice or murine cells deficient for ligase IV have a substantial reduction in CSR (Yan *et al.*, 2007; Han and Yu, 2008). Knock-out murine cells for XRCC4 also have a substantial reduction in CSR (Soulas-Sprauel *et al.*, 2007). Hence, all studies point toward the XRCC4:DNA ligase IV complex as being of major importance for CSR. However, it is uncertain how much non-NHEJ components or pathways can compensate when a complete NHEJ system is not present. This is discussed under the *Alternative NHEJ* section below.

5.3.5 *Terminal microhomology usage in NHEJ*

About 20–50% of NHEJ events are found to occur at one or more nucleotides of shared microhomology at the terminus of each DNA end. Joining in mammalian cells can occur between two DNA ends that share no microhomology. For example, using extrachromosomal DNA substrates for V(D)J recombination, one can engineer the two coding ends to have no shared microhomology. Yet overall, V(D)J recombination is not reduced in frequency. NHEJ might not be the rate-limiting step for recombination of those extrachromosomal substrates, but such studies suggest that lack of microhomology is not an absolute requirement.

5.3.6 *Alternative NHEJ*

Murine and yeast genetic studies have shown that some level of end joining can occur in the absence of ligase IV (Daley *et al.*, 2005; Yan *et al.*, 2007; Han and Yu, 2008; Soulas-Sprauel *et al.*, 2007). Because the only remaining ligase activity in the cells is due to ligase I in *S. cerevisiae* or ligase I or III in vertebrate cells, residual end-joining in the absence of ligase VI must be completed by ligase I or III. Most of these joining events rely more heavily on the use of terminal microhomology than NHEJ in wild-type cells.

In *in vitro* systems using purified NHEJ proteins, human ligase I and III are able to join DNA ends that are not fully compatible (e.g. joining across gaps in the ligated strand), though it is still substantially less efficient than joining by the XRCC4:ligase IV complex (Gu *et al.*, 2007a; Gu *et al.*, 2007b). This joining by ligase I or III is a bit more efficient with two or more base pairs of terminal microhomology to stabilize the ends. Therefore, in the absence of the ligase IV complex, the peak microhomology use increases from zero to 2.5bp.

Regarding the relative ratio of ligase IV-mediated CSR to non-ligase IV-mediated joining, two measurements have been made in murine cells. In one study, cells lacking ligase IV are removed from mice and stimulated to undergo CSR (Yan *et al.*, 2007). Measurements of switch recombination can be done as early as 60 hrs after stimulation and show that end joining without ligase IV is reduced ~2.5–fold. In another study, knock-out of ligase IV was done in a murine cell line and measurements could be done as early as 24 hrs, at which time the joining without ligase IV was reduced approximately nine-fold (Han and Yu, 2008). In both cases, the joining is almost certainly catalyzed by ligase I or III. The second study suggests that the joining by ligase I or III is substantially less efficient at early times. In both studies, given sufficient time, the joining by ligase I or III improves to about half of that of the wild-type cells. Additionally, end-joining in wild-type cells was much less dependent on terminal microhomology than joining in the ligase IV knock-out cells. One reasonable interpretation of these two studies is that ligase IV is more efficient (and perhaps faster) at joining incompatible DNA ends *in vivo*; however, ligase I or III can join ends at a lower efficiency, especially when terminal microhomology can stabilize the DNA ends.

Ku-independent end joining has been observed in mammalian cells (Weinstock *et al.*, 2007). In vertebrate V(D)J recombination, when the two DNA ends share four base pairs of terminal microhomology, the dependence on Ku for joining efficiency can be only 2.5–fold (Weinstock *et al.*, 2007). Hence, terminal microhomology can substitute for the presence of Ku. Thus, when Ku is absent, terminal microhomology may provide end-stability, consistent with observations in biochemical systems (Gu *et al.*, 2007b).

5.4 Concluding Comments and Future Questions

A challenge in the field is to understand the mechanism by which AID-dependent DNA double strand breaks are generated in switch regions: once such breaks are created (see Section 2 of this chapter), the rejoining of the DNA ends by NHEJ or by alternative enzymes (see Section 3.6 of this chapter) likely proceeds through well-studied mechanisms. Therefore, it will be important to clarify the role of MMR proteins in the generation of DNA double-stranded breaks, in the context of repetitive regions. The role of R-loops is also of substantial interest, particularly in light of the apparent reliance of the local chromatin structure of class switch zones on the R-loop-forming regions (Rajagopal *et al.*, 2009; Wang *et al.*, 2009); as is the contribution of AID motifs (WRC) especially when counterbalanced by the G-density of class switch repeats. Finally it remains important to understand the relative contribution of NHEJ versus additional components in the synapsis and rejoining of the double-strand breaks. For example, what is the role of the RAD50:MRE11:NBS1 (MRN) complex in CSR? Is it direct or indirect? Is it involved in all CSR events, or only a subset? Does it serve a synaptic function, and if so, it is the sole synaptic complex or are there additional proteins that can play that role? These are all interesting and important questions not only for the CSR field but also for the field of DNA repair more generally.

5.5 References

1. Bardwell P.D., Woo C.J., Wei K. *et al.* (2004). Altered somatic hypermutation and reduced class-switch recombination in exonuclease 1-mutant mice. *Nat Immunol* **5**: 224–229.
2. Begum N.A., Kinoshita K., Kakazu N. *et al.* (2004). Uracil DNA glycosylase activity is dispensable for immunoglobulin class switch. *Science* **305**: 1160–1163.

3. Bosma G.C., Kim J., Urich T. *et al.* (2002). DNA-dependent protein kinase activity is not required for immunoglobulin class switching. *J Exp Med* **196**: 1483–1495.
4. Bransteitter R., Pham P., Scharff M.D. *et al.* (2003). Activation-induced cytidine deaminase deaminates deoxycytidine on single-stranded DNA but requires the action of RNase. *Proc Natl Acad Sci USA* **100**: 4102–4107.
5. Casellas R., Nussenzweig A., Wuerffel R. *et al.* (1998). Ku80 is required for immunoglobulin isotype switching. *EMBO J* **17**: 2404–2411.
6. Daley J.M., Laan R.L., Suresh A. *et al.* (2005). DNA joint dependence of pol X family polymerase action in nonhomologous end joining. *J Biol Chem* **280**: 29030–29037.
7. Demple B., Harrison L. (1994). Repair of oxidative damage to DNA: enzymology and biology. *Annu Rev Biochem* **63**: 915–948.
8. Di Noia J.M., Neuberger M.S. (2002). Altering the pathway of immunoglobulin hypermutation by inhibiting uracil-DNA glycosylase. *Nature* **419**: 43–48.
9. Di Noia J.M., Neuberger M.S. (2007). Molecular mechanisms of antibody somatic hypermutation. *Annu Rev Biochem* **76**: 1–22.
10. Di Noia J.M., Williams G.T., Chan D.T. *et al.* (2007). Dependence of antibody gene diversification on uracil excision. *J Exp Med* **204**: 3209–3219.
11. Dickerson S.K., Market E., Besmer E. *et al.* (2003). AID mediates hypermutation by deaminating single stranded DNA. *J Exp Med* **197**: 1291–1296.
12. Eccleston J., Schrader C.E., Yuan K. *et al.* (2009). Class switch recombination efficiency and junction microhomology patterns in Msh2-, Mlh1-, and Exo1-deficient mice depend on the presence of mu switch region tandem repeats. *J Immunol* **183**: 1222–1228.
13. Ehrenstein M.R., Neuberger M.S. (1999). Deficiency in Msh2 affects the efficiency and local sequence specificity of immunoglobulin class-switch recombination: parallels with somatic hypermutation. *EMBO J* **18**: 3484–3490.
14. Franco S., Murphy M.M., Li G. *et al.* (2008). DNA-PKcs and Artemis function in the end-joining phase of immunoglobulin heavy chain class switch recombination. *J Exp Med* **205**: 557–564.
15. Gu J., Lu H., Tippin B. *et al.* (2007a). XRCC4:DNA ligase IV can ligate incompatible DNA ends and can ligate across gaps. *EMBO J* **26**: 1010–1023.
16. Gu J., Lu H., Tsai A.G., Schwarz K. *et al.* (2007b). Single-stranded DNA ligation and XLF-stimulated incompatible DNA end ligation by the XRCC4-DNA ligase IV complex: influence of terminal DNA sequence. *Nucleic Acids Res* **35**: 5755–5762.
17. Guikema J.E., Linehan E.K., Tsuchimoto D. *et al.* (2007). APE1- and APE2-dependent DNA breaks in immunoglobulin class switch recombination. *J Exp Med* **204**: 3017–3026.
18. Han L., Yu K. (2008). Altered kinetics of nonhomologous end joining and class switch recombination in ligase IV-deficient B cells. *J Exp Med* **205**: 2745–2753.
19. Imai K., Slupphaug G., Lee W.I. *et al.* (2003). Human uracil-DNA glycosylase deficiency associated with profoundly impaired immunoglobulin class-switch recombination. *Nat Immunol* **4**: 1023–1028.

20. Li Z., Scherer S.J., Ronai D. *et al.* (2004). Examination of Msh6- and Msh3-deficient mice in class switching reveals overlapping and distinct roles of MutS homologues in antibody diversification. *J Exp Med* **200**: 47–59.
21. Luby T.M., Schrader C.E., Stavnezer J. *et al.* (2001). The mu switch region tandem repeats are important, but not required, for antibody class switch recombination. *J Exp Med* **193**: 159–168.
22. Manis J.P., Gu Y., Lansford R. *et al.* (1998). Ku70 is required for late B cell development and immunoglobulin heavy chain class switching. *J Exp Med* **187**: 2081–2089.
23. Martin A., Li Z., Lin D.P. *et al.* (2003). Msh2 ATPase activity is essential for somatic hypermutation at a-T basepairs and for efficient class switch recombination. *J Exp Med* **198**: 1171–1178.
24. Min I.M., Rothlein L.R., Schrader C.E. *et al.* (2005). Shifts in targeting of class switch recombination sites in mice that lack mu switch region tandem repeats or Msh2. *J Exp Med* **201**: 1885–1890.
25. Min I.M., Schrader C.E., Vardo J. *et al.* (2003). The Smu tandem repeat region is critical for Ig isotype switching in the absence of Msh2. *Immunity* **19**: 515–524.
26. Neuberger M.S., Di Noia J.M., Beale R.C. *et al.* (2005). Somatic hypermutation at A.T pairs: polymerase error versus dUTP incorporation. *Nat Rev Immunol* **5**: 171–178.
27. Peron S., Metin A., Gardes P. *et al.* (2008). Human PMS2 deficiency is associated with impaired immunoglobulin class switch recombination. *J Exp Med* **205**: 2465–2472.
28. Petersen-Mahrt S.K., Harris R.S., Neuberger M.S. (2002). AID mutates E. coli suggesting a DNA deamination mechanism for antibody diversification. *Nature* **418**: 99–103.
29. Pham P., Bransteitter R., Petruska J. *et al.* (2003). Processive AID-catalysed cytosine deamination on single-stranded DNA simulates somatic hypermutation. *Nature* **424**: 103–107.
30. Rada C., Di Noia J.M., Neuberger M.S. (2004). Mismatch recognition and uracil excision provide complementary paths to both Ig switching and the A/T-focused phase of somatic mutation. *Mol Cell* **16**: 163–171.
31. Rada C., Williams G.T., Nilsen H. *et al.* (2002). Immunoglobulin isotype switching is inhibited and somatic hypermutation perturbed in UNG-deficient mice. *Curr Biol* **12**: 1748–1755.
32. Rajagopal D., Maul R.W., Ghosh A. *et al.* (2009). Immunoglobulin switch mu sequence causes RNA polymerase II accumulation and reduces dA hypermutation. *J Exp Med* **206**: 1237–1244.
33. Riballo E., Kuhne M., Rief N. *et al.* (2004). A pathway of double-strand break rejoining dependent upon ATM, Artemis, and proteins locating to gamma-H2AX foci. *Mol Cell* **16**: 715–724.
34. Rooney S., Alt F.W., Sekiguchi J. *et al.* (2005). Artemis-independent functions of DNA-dependent protein kinase in Ig heavy chain class switch recombination and development. *Proc Natl Acad Sci USA* **102**: 2471–2475.

35. Sabouri Z., Okazaki I.M., Shinkura R. *et al.* (2009). Apex2 is required for efficient somatic hypermutation but not for class switch recombination of immunoglobulin genes. *Int Immunol* **21**: 947–955.
36. Schrader C.E., Edelmann W., Kucherlapati R. *et al.* (1999). Reduced isotype switching in splenic B cells from mice deficient in mismatch repair enzymes. *J Exp Med* **190**: 323–330.
37. Soulas-Sprauel P., Le Guyader G., Rivera-Munoz P. *et al.* (2007). Role for DNA repair factor XRCC4 in immunoglobulin class switch recombination. *J Exp Med* **204**: 1717–1727.
38. Stavnezer J., Schrader C.E. (2006). Mismatch repair converts AID-instigated nicks to double-strand breaks for antibody class-switch recombination. *Trends Genet* **22**: 23–28.
39. Vora K.A., Tumas-Brundage K.M., Lentz V.M. *et al.* (1999). Severe attenuation of the B cell immune response in Msh2-deficient mice. *J Exp Med* **189**: 471–482.
40. Wang L., Wuerffel R., Feldman S. *et al.* (2009). S region sequence, RNA polymerase II, and histone modifications create chromatin accessibility during class switch recombination. *J Exp Med* **206**: 1817–1830.
41. Weinstock D.M., Brunet E., Jasin M. (2007). Formation of NHEJ-derived reciprocal chromosomal translocations does not require Ku70. *Nat Cell Biol* **9**: 978–981.
42. Xue K., Rada C., Neuberger M.S. (2006). The in vivo pattern of AID targeting to immunoglobulin switch regions deduced from mutation spectra in msh2-/- ung-/- mice. *J Exp Med* **203**: 2085–2094.
43. Yan C.T., Boboila C., Souza E.K. *et al.* (2007). IgH class switching and translocations use a robust non-classical end-joining pathway. *Nature* **449**: 478–482.
44. Yu K., Chedin F., Hsieh C.L. *et al.* (2003). R-loops at immunoglobulin class switch regions in the chromosomes of stimulated B cells. *Nat Immunol* **4**: 442–451.
45. Yu K., Huang F.T., Lieber M.R. (2004). DNA substrate length and surrounding sequence affect the activation-induced deaminase activity at cytidine. *J Biol Chem* **279**: 6496–6500.
46. Zarrin A.A., Alt F.W., Chaudhuri J. *et al.* (2004). An evolutionarily conserved target motif for immunoglobulin class-switch recombination. *Nat Immunol* **5**: 1275–1281.

Chapter 6

Error-Prone and Error-Free Resolution of AID Lesions in SHM

Peter H.L. Krijger,[1] Ursula Storb,[2,3] and Heinz Jacobs[1]

[1]Division of Immunology, The Netherlands Cancer Institute
1066 CX, Amsterdam, The Netherlands
E-mail: h.jacobs@nki.nl

[2]Committee on Immunology, University of Chicago
Chicago, IL 60637

[3]Department of Molecular Genetics and Cell Biology
University of Chicago, Chicago, IL 60637
E-mail: stor@uchicago.edu

Previous chapters outline present knowledge on *cis*-acting elements and *trans*-acting proteins that target AID to Ig loci. This chapter reveals the astonishing creativity and enormous flexibility of nature in transforming evolutionary highly conserved DNA repair and DNA damage-tolerance pathways into highly effective DNA mutation pathways. The goal is to create a virtually unlimited antibody repertoire from within the confines of a size-limited genome. The trick is to improve adaptive B cell immunity by introducing point mutations into the variable region of Ig genes in antigen-activated B cells. The result is a highly diversified secondary antibody repertoire which, because of point mutations in the antigen-binding regions CDR1, CDR2 and CDR3 and B cell selection, is of higher affinity. As noted in the first chapter, this process of SHM takes place in antigen-activated B cells of the germinal center, a defined histological structure that forms transiently in secondary lymphoid tissue during immune responses.

6.1 Introduction

SHM is initiated by the activation-induced cytidine deaminase (AID), an enzyme found to be differentially expressed in B cells of the GC (Muramatsu *et al.*, 2000). AID deaminates cytosine (C) to uracil (U) within single-stranded DNA (Pham *et al.*, 2003; Chaudhuri *et al.*, 2003; Dickerson *et al.*, 2003; Ramiro *et al.*, 2003), and consistent with earlier observations (Dorner *et al.*, 1998; Milstein *et al.*, 1998), targets both DNA strands in the variable regions of Ig genes. While the primary lesion is restricted to cytosine deamination, SHM occurs equally efficient at G/C and A/T base pairs. During the last decennium, humans, mice and cell lines carrying defined genetic alterations in DNA repair and DNA damage-tolerance elements have revealed detailed insights into the molecular pathways controlling the generation of defined point mutations in hypermutating Ig genes. The combination of these pathways enables hypermutating B cells to generate the entire spectrum of nucleotide substitutions at a rate of one per thousand bases per generation, several orders of magnitude greater than spontaneous mutagenesis. To establish these point mutations at and around the initial U lesion, B cells proved to be highly creative in transforming established DNA repair pathways into effective mutator pathways. The mutagenic transformation of these faithful repair pathways appears to depend strongly on a family of specialized DNA polymerases with defined error signatures. To date, three main mutation pathways have been identified that contribute to the mutagenic processing of U generated by AID (Fig. 6.1a). This chapter focuses on the role of these pathways and presents knowledge/models regarding their (in)dependence in establishing specific point mutations. In addition, we focus on the regulatory aspects controlling the establishment of defined mutations.

6.2 Direct Replication Across the Uracil: G/C Transitions

Besides intentional cytosine deamination of Ig genes by AID in hypermutating B cells, spontaneous cytosine deamination or UMP incorporation occurs frequently and is one of the most common lesions in our genome. If not removed from the genome in a timely fashion, a

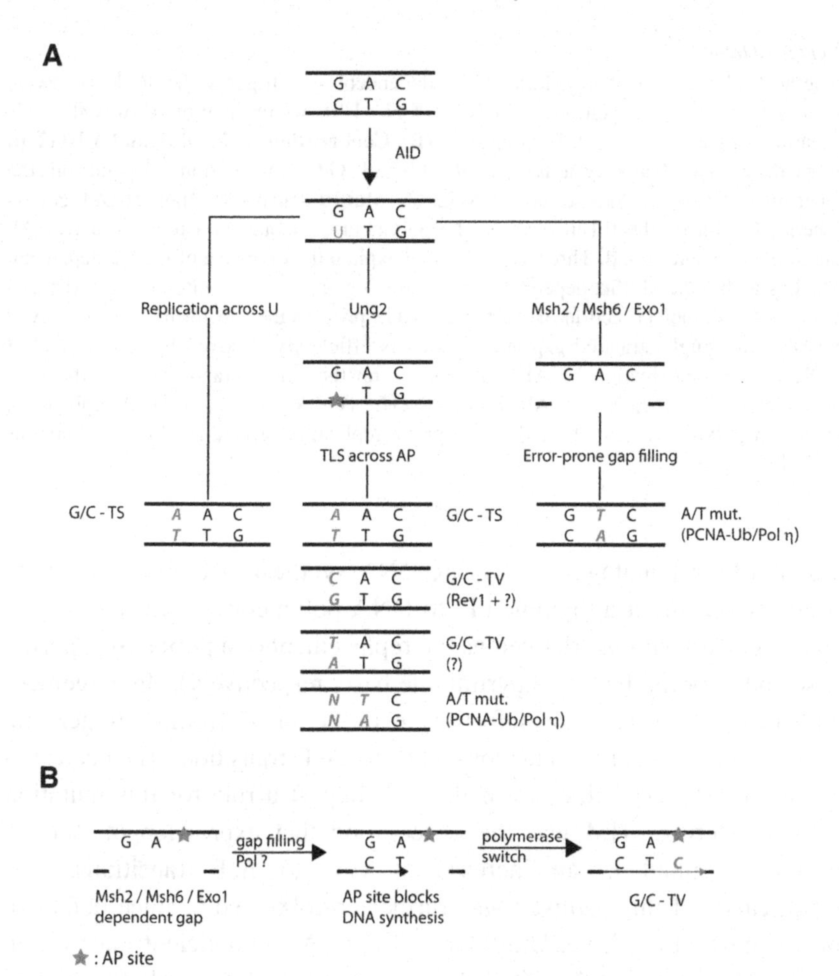

Figure 6.1. **(A) The three pathways of SHM downstream of AID.** AID deaminates C to U. Three error-prone repair pathways can process the U: I) Left panel: Direct replication across U; during replication a U in the template strand will mimic a template T and generate G/C to A/T transitions (TS). II) Middle panel: UNG2-dependent SHM; upon removal of the U by UNG2 an abasic site (AP site, indicated by a star) is generated. Replication across this UNG2-dependent, non-instructive AP site generates G/C transitions and transversions (TV). Rev1 generates G to C transversions. Other unknown polymerases (?) generate G/C transversions. A minority of A/T mutations (~10%) is generated downstream of UNG2. A/T mutations strongly depend on the synergistic action of Polη and its preferred DNA sliding clamp, the monoubiquitylated PCNA (PCNA-Ub). III) Right panel: MutSα-dependent SHM; the U/G mismatch generated by AID can also be recognized by MutSα. MutSα, an unknown 5′ endonuclease and Exo1

(Continued on p. 100)

(*Continued from p. 99*)
generate a large gap. This MutSα/Exo1-dependent gap triggers the Rad6 pathway, leading to monoubiquitylation of PCNA (PCNA-Ub), which in turn recruits the A/T mutator Polη to generate A/T mutations. **(B) Cooperation of MutSα and UNG2 in generating G/C transversions.** About 50% of G/C transversions depend on the synergistic action of MutSα and UNG2. A MutSα-dependent single-strand gap is generated. During the fill-in reaction the polymerase encounters a non-instructive AP site in the template strand. Three scenarios can explain the existence of UNG2-dependent AP sites within the MutSα-dependent single-strand gap: 1) The AP site may preexist as a result of the combined action of AID and UNG2 prior to the gap formation; or 2) A U exists in the single-stranded gap and as such is efficiently removed by UNG2; and/or 3) Secondary deamination by AID takes place on the single-strand gap, and the U is immediately processed into an AP site by UNG2. This AP site blocks DNA replication, induces a polymerase switch, and error- prone replication generates G/C transversions (G/C-TV).

uracil is highly mutagenic. During DNA synthesis a U in the template strand will instruct a thymine (T) to DNA polymerases, causing G to A and C to T transitions (defined as the replacement of a purine by a purine base and a pyrimidine by a pyrimidine base, respectively). In agreement with this notion, bacteria defective in removing U from their genome have a high rate of spontaneous G/C to A/T transitions (Duncan and Miller, 1980). The first set of data to suggest a role for this mutation pathway during SHM was the observation that expression of AID in bacteria resulted in an increase of G/C to A/T transitions. The significance of this pathway was further corroborated in mice defective in the removal of U, i.e. Ung/Msh2- or Ung/Msh6-deficient mice (Shen *et al.*, 2006; Rada *et al.*, 2004). In these mice, the base exchange pattern of mutated Ig genes was extremely compromised, showing only G/C transitions. These data clearly revealed a major pathway in the generation of G/C transition mutations and strongly suggested that the remaining point mutations of the hypermutation spectrum require further modifications of the initial U lesion (Fig. 6.1a, left panel). In fact, as outlined below, SHM critically depends on two generic DNA repair factors, capable of recognizing U in the DNA, the base excision repair (BER) factor UNG2 and the mismatch repair (MMR) factor MutSα (Fig. 6.1a, middle and right panel, respectively).

6.3 UNG2-Dependent SHM Across AP Sites: G/C Transversions and Transitions

As mentioned, spontaneous cytosine deamination is one of the most frequent DNA lesions. To maintain the integrity of the genome, U and other highly mutagenic base modifications can effectively be removed by a multistep repair process, known as BER. BER is initiated by a family of highly efficient, partially redundant DNA-glycosylases capable of recognizing and removing modified bases from our genome. DNA-glycosylases catalyze the hydrolysis of the N-glycosidic bond that links the base to the deoxyribose-phosphate backbone. After the base excision step, the DNA duplex now harbors an apurinic or apyrimidinic site in its backbone (AP: a site without a purine or pyrimidine base analogue); this is also known as an abasic site. The repair of AP sites, the common product of glycosylase action, requires a second class of BER enzymes known as AP endonucleases (APE1 and APE2 in mammals), which generate nicks in the duplex DNA by hydrolysis of the phosphodiester bond immediately 5′ to the AP site. In mammalian cells, further processing involves DNA polymerase β (Polβ). Polβ has two enzymatic activities, a large C-terminal DNA polymerase domain and a small N-terminal DNA-deoxy-ribophosphodiesterase (dRpase). While the dRpase activity of Polβ removes the dRp nicked by APE, the polymerase activity fills up the single nucleotide gap. While the lack of an intrinsic proofread activity renders Polβ error-prone (1–2 misinsertions per 10000), accuracy might be gained at the level of DNA ligation and postreplicative MMR (Kunkel, 1985; Friedberg *et al.*, 2006).

In addition to this short patch BER pathway (1nt), the nature of the DNA glycosylase and/or lesion may require an alternative pathway of BER, known as long-patch BER. Long-patch BER involves the flap-endonuclease 1 (FEN1), which after displacement of the strand containing the modified base (flap) makes an incision to generate a long single-strand patch (2–8nt). Repair synthesis of this long patch requires components of the replication machinery such as Polδ and Polε, the DNA sliding clamp proliferating cell nuclear antigen (PCNA) as well as DNA ligase 1. Given the accuracy of Polδ, and Polε, long-patch BER is effective in maintaining genome integrity after base damage. Besides replicative DNA polymerases, Polβ may also contribute to long-patch BER (Podlutsky *et al.*, 2001; Singhal and Wilson, 1993).

In mammals, four DNA glycosylases have been identified that can hydrolyze U from the DNA backbone: Uracil-DNA glycosylase (UNG); SMUG DNA glycosylase (SMUG1); methyl-binding domain glycosylase 4 (MBD4); and thymine DNA glycosylase (TDG) (Krokan *et al.*, 2002). Although redundant in their enzymatic activity *in vitro*, only UNG appears to be essential during SHM (Bardwell *et al.*, 2003; Di Noia *et al.*, 2006; Rada *et al.*, 2002; Rada *et al.*, 2004). Two alternative splice variants of UNG exist, a mitochondrial UNG1 and a nuclear UNG2. *Ung* mutant B cells lack most G/C transversions (Rada *et al.*, 2002). These transversions do not depend on Polβ, as Polβ-deficient B cells mutate normally (Esposito *et al.*, 2000). The role of the APE endonucleases during SHM is unknown, as APE1 deficiency causes early embryonic lethality (Xanthoudakis *et al.*, 1996). Most likely, during SHM, AP sites are not removed prior to replication or repair synthesis. As AP sites are non-instructive they cause replicative DNA polymerases to stall. To continue DNA synthesis across an AP site, specialized TLS polymerases are recruited (see below) that tolerate such blocking lesions and thereby generate G/C transversion as well as transition mutations (Fig. 6.1a, middle panel). Besides APE the Mre11/Rad50/Nbs1 (MRN) complex has been proposed to initiate mutagenesis by cleaving AP sites (Larson *et al.*, 2005; Yabuki *et al.*, 2005). As the resulting ends cannot be extended by high-fidelity DNA polymerases, it has been suggested that error-prone DNA polymerases take over to introduce mutations when filling the gap. Future studies in MRN-deficient cells should clarify the relevance of this pathway in SHM.

6.4 MutSα-Dependent SHM at MMR Gaps: A/T Mutations

Cytosine deamination in the DNA helix generates a U:G mismatch. Besides BER, the U:G mismatch can be processed by DNA mismatch repair (MMR; Wilson *et al.*, 2005; Schanz *et al.*, 2009). MMR is an evolutionarily-conserved process that corrects mismatches that have escaped proofreading during DNA replication. The MMR process involves a complex interplay of MMR-specific proteins with the replication and/or recombination machinery (Jiricny, 1998). MMR is initiated by the binding of the mismatch-recognition factors, MutSα (MSH2/MSH6 complex) to single base mismatches or MutSβ (MSH2/MSH3 complex) to insertion/deletion loops that arise during recombination or from errors of DNA polymerases. Mammalian MMR is proposed to initiate at strand discontinuities, such as nicks or gaps that are distal to the mispair (Modrich, 2006; Schanz *et al.*, 2009). The recruitment of MutL homologues (MutLα: MLH1-PMS2 complex; MutLβ: MLH1-PMS1 complex) stabilizes the mismatch-bound MutSα complex and appears to prohibit sliding of MutS. Exonuclease-1 (Exo1) mediated degradation of the error-containing strand depends on an incision 5' of the mismatch. This incision may involve the nuclease activity of PMS2 (Kadyrov *et al.*, 2006) or an alternative nuclease. Once the mismatch is removed, resynthesis of the degraded region by a DNA polymerase, followed by sealing of the remaining nick by DNA ligase, completes the repair process.

Remarkably, given the protective nature of MMR in preventing mutations arising from mismatched non-Watson-Crick base pairs, early studies in mismatch repair mutant mice revealed a selective role of the mismatch recognition complex MutSα as well as Exo1 in establishing somatic mutation at template A/T around the initial U:G mismatch. Interestingly, while MSH2, MSH6 and Exo-1-deficient B cells lack 80–90% of all A/T mutations, the SHM phenotype appears less pronounced or even normal in B cells lacking other MMR components such as PMS2, MLH1, MLH3 and MSH3 (Rada *et al.*, 1998; Wiesendanger *et al.*, 2000; Bardwell *et al.*, 2004; Martomo *et al.*, 2004; Phung *et al.*, 1999; Phung *et al.*, 1998; Ehrenstein *et al.*, 2001; Winter *et al.*, 1998; Jacobs *et al.*, 1998; Frey *et al.*, 1998). These data suggest that during SHM, selective components of the mismatch repair machinery are required to generate a single-strand gap. In contrast to conventional, postreplicative MMR, the gap-filling process during SHM appears to employ error-prone TLS polymerase(s) that generate predominantly A/T mutations (Fig. 6.1a, right panel). At present the identity of the incision maker 5′ to the mismatch, which is required for Exo1 is unknown.

UNG2/APE has been proposed (Schanz *et al.*, 2009), but given the fact that A/T mutations are unaffected in UNG-deficient B cells (Rada *et al.*, 2002; Krijger *et al.*, 2009), alternative uracil glycosylases might take over.

6.5 UNG-Dependent A/T Mutations

A significant proportion of A/T mutations (10–20%) are found in MSH2-deficient GC B cells, but not in UNG2/MSH2 double-deficient GC B cells, indicating that UNG2-dependent mutagenesis generates the above mentioned proportion of A/T mutations (Rada *et al.*, 2004). Whether UNG2-dependent A/T mutations are generated during long-patch BER, i.e. within the strand containing the AP site, or alternatively during the extension phase of TLS across the AP site, is currently unknown. Mice deficient for FEN1 are embryonic lethal, and analysis of B cells from mice expressing a hypomorph variant of FEN1 showed no SHM phenotype (Larsen *et al.*, 2008): conditional knock-out mice or deficient cell lines will have to be generated to determine whether this pathway contributes to SHM.

6.6 Half of all G/C Transversions Require MutSα and UNG2

As mentioned, G/C transversions strongly depend on AP sites generated by UNG2, while most A/T mutations depend on MutSα. These data implicated a strict separation between these pathways in establishing defined mutations. However, a quantitative analysis indicated that approximately 50% of all G/C transversions requires the combined activity of MutSα and UNG2 (Krijger *et al.*, 2009). These data suggest a model in which DNA polymerases involved in the gap-filling process downstream of MutSα become stalled by AP sites. The subsequent activation of specialized TLS polymerases enables TLS across the non-instructive AP site, thereby generating MutSα/UNG2-dependent G/C transversions (Fig. 6.1b). Three scenarios can explain the existence of

UNG2-dependent AP sites within the MutSα/Exo-1- dependent single-strand gap: the AP site may preexist as a result of the combined action of AID and UNG2 prior to gap formation; or a U exists in the single-stranded gap and as such is efficiently removed by UNG2; and/or secondary deamination by AID takes place on the single-strand gap, and the U is immediately processed into an AP site by UNG2. Further studies should reveal which of the above sources of AP substrates contribute to the generation of MutSα/UNG2-dependent G/C transversions.

6.7 Translesion Synthesis DNA Polymerases

To explain the unusually high mutation rate of SHM, error-prone polymerases were postulated about half a century ago (Brenner and Milstein, 1966). Yet, only during the past two decades were the existence of error-prone TLS DNA polymerases revealed. Their characterization *in vitro* and *in vivo* indicated an error rate that easily matches the one of SHM. The largest family, the Y-family of TLS polymerases comprising Polη, Polι, Polκ and Rev1 is characterized by five highly conserved motifs located in the catalytic domain (Prakash *et al.*, 2005). Other DNA polymerases like the B-family member Polζ (Gan *et al.*, 2008) and the A-family member Polθ (Seki *et al.*, 2004), as well as others, display TLS activity. TLS polymerases share the unique capacity to bypass DNA lesions, i.e. they can continue replication in the presence of noninstructive or misinstructive DNA lesion that otherwise may stall the replicative Polε and Polδ. In general, TLS is thought to proceed in a two-step mode (Shachar *et al.*, 2009; Johnson *et al.*, 2000; Ziv *et al.*, 2009): (1) Incorporation of nucleotide(s) directly opposite of the lesion. (2) Elongation from the distorted or bulky non-Watson-Crick base pairs by an extender TLS polymerase. A prerequisite for TLS is the lack of proofreading activity. Indeed, in contrast to replicative DNA polymerases, TLS polymerases lack proofreading activity. The capacity of TLS polymerases to accommodate non Watson-Crick base pairs within their catalytic center is beneficial and accurate when replication across modified bases is required (such as UV-C-induced cyclic pyrimidine dimers (CPD) by Polη). Polη has the unique capacity to insert and accommodate two A opposite a *cis*-syn TT-CPD, thereby maintaining the genetic information of the newly synthesized DNA in the presence of a damaged template (Johnson *et al.*, 1999). As shown *in vitro* and *in vivo* Polζ and Polκ are efficient in extending from these bypassed CPD lesions (Washington *et al.*, 2002). Once extended, proofreading-proficient high-fidelity DNA polymerase cannot detect the lesion any longer and resumes DNA synthesis. However, TLS polymerases can become highly mutagenic when replicating across undamaged DNA and defined lesions such as AP sites (Jansen *et al.*, 2007; Prakash *et al.*, 2005). Since each polymerase displays its own mutagenic signature, alterations in the mutation spectrum can often be attributed retrospectively to the absence of, or failure in, activating specific polymerases. This preference has been useful in the identification of DNA polymerases involved in SHM.

6.7.1 *Polη generates most A/T mutations*

Polη, a polymerase that is absent or hypomorphic in patients with the variant form of Xeroderma Pigmentosum (XP-V; Johnson *et al.*, 1999; Masutani *et al.*, 1999), was the first to be linked to SHM. B cells from these patients showed an altered spectrum of somatic point mutations (Zeng *et al.*, 2001). A significant reduction in mutations at A/T base pairs was associated with a relative increase of mutations at template G/C. These observations were confirmed in mouse models defective for Polη (Delbos *et al.*, 2005; Martomo *et al.*, 2005). Consistent with these *in vivo* data, Polη has a preference to insert mismatched nucleotides opposite template T (Rogozin *et al.*, 2001) but is ineffective in handling AP sites (Haracska *et al.*, 2001) *in vitro*. Apparently, Polη is required in generating most A/T mutations, a phenotype closely resembling MutSα-deficient B cells. These data suggest that Polη is employed mainly downstream of MSH2. In addition to its role downstream of MutSα, Pol η is responsible for the remaining A/T mutations downstream of UNG2, as deduced from SHM analysis in MSH2 and MSH2/Polη-deficient mice (Delbos *et al.*, 2007). Although postulated for a long time

(Brenner and Milstein, 1966) these observations provided the first evidence for the existence and involvement of error-prone DNA polymerases in establishing defined point mutations in hypermutated Ig genes and stimulated efforts to identify other TLS polymerases involved in this process.

6.7.2 *Polκ can partially compensate for Polη deficiency*

Polκ seems inessential for somatic hypermutation as demonstrated independently in Polκ-deficient mice (Schenten *et al.*, 2002; Shimizu *et al.*, 2003). However, the residual A/T mutations found in Polη-deficient B cells have been demonstrated to depend on Polκ and at least a third yet unidentified polymerase (Faili *et al.*, 2009). This observation is compatible with the error-signature of Polκ *in vitro* (Ohashi *et al.*, 2000). Apparently, Polκ can substitute Polη whereas other polymerases of the Y-family, for example Rev1, cannot, as revealed by the normal generation of G to C transversions in Polη-deficient mice (see below).

6.7.3 *TLS polymerase Rev1 generates G to C transversions*

Rev1 is selective in its nucleotide incorporation activity as it only incorporates dCMP and therefore in its strictest sense should be regarded as a deoxycytidyl transferase rather than a *bona fide* DNA polymerase. *In vitro*, Rev1 is capable of bypassing both uracil residues and AP sites (Nelson *et al.*, 1996). Rev1 harbors a BRCA1 C-terminal (BRCT) domain in its N-terminus (Gerlach *et al.*, 1999). The BRCT domain of Rev1 was shown to regulate TLS of AP sites in yeast (Haracska *et al.*, 2001). However, hypermutated Ig genes of memory B cells derived from Rev1 mutant mice lacking the N-terminal BRCT domain, revealed no changes in the base exchange pattern. This indicates that the BRCT domain is dispensable in establishing somatic mutations in the V regions of Ig genes (Jansen *et al.*, 2005), leaving the possibility that the catalytic domain of Rev1 might play a role in SHM. Indeed, B cells derived from Rev1-deficient mice as well as chicken DT40 cells exhibited an altered base exchange pattern (Jansen *et al.*, 2006; Ross and Sale, 2006). In agreement with the reported *in vitro* ability of Rev1 to bypass AP sites by incorporating cytosine residues opposite of this lesion, C to G and G to C transversions were significantly reduced in the absence of Rev1, but the remaining C to G and G to C transversions indicate that other polymerases can make these transversions. This reduction was associated with a relative increase in A to T, C to A and T to C mutations. In the presence of Rev1, an AP site – derived from cytosine deamination by AID followed by the removal of the uracil by UNG2 during SHM – will be bypassed by the incorporation of a cytidine residue. In the absence of Rev1 however, other TLS polymerases with a distinct mutation signature are likely to bypass this lesion, thereby favoring the introduction of other mutations.

6.7.4 *Polι, a story to be finished*

In vitro Polι prefers to insert a G rather than an A opposite of T (Zhang *et al.*, 2000; Tissier *et al.*, 2000), which could also explain the increase in T to C transitions seen in the absence of Rev1. In addition, Polι has a preference to insert either G or T residues opposite of AP sites (Zhang *et al.*, 2001). While incorporation of G opposite of an AP site will faithfully restore the initial AID-induced lesion, the introduction of a T will result in C to A and G to T transversions. Actually, the TLS polymerase(s) involved in establishing these transversions during SHM remain to be identified. Remarkably, no changes in SHM were observed in B cells derived from a *129/J*-mouse strain that carries a spontaneous nonsense mutation in the Polι gene (McDonald *et al.*, 2003). Western blot analysis on testis extracts showed the absence of Polι in this strain. Nevertheless, it has been noted that there may be tissue- specific and functional alternative splice forms of Polι, and 'Polι activity' seems to be retained in brain extract from this mouse strain (Gening and Tarantul, 2006). In this context, *129/J*-derived B cells should be tested for the presence of hypomorph versions of Polι. At present, one cannot formally exclude the possibility that Polι is involved in SHM. Analysis of B cells derived from mouse mutants carrying a targeted deletion of Polι will resolve this issue.

6.7.5 *Polζ, an extender polymerase that might be replaceable*

Polζ is a heterodimer composed of a catalytic Rev3 and structural Rev7 protein that extends efficiently from mispaired primer termini on undamaged DNA. Rev3 deficiency leads to embryonic lethality, suggesting a critical role of this DNA polymerase in TLS (Gan *et al.*, 2008). Rev3-deficient chicken DT40 B cells revealed a central role of Rev3 in maintaining genome stability. Besides its critical role in TLS, Rev3-deficient cells showed reduced gene- targeting efficiencies and a significant increase in the level of genomic breaks after ionizing radiation (Sonoda *et al.*, 2003). Similarly, cell lines established from Rev3-deficient mice are genetically unstable and prone to apoptosis (Jansen *et al.*, 2009). Consistent with the extension function of Polζ from mispaired primer templates, a knock-down of the catalytic subunit Rev3 by antisense oligos in human B cells or antisense RNA in transgenic mice revealed a decrease in the frequency of somatic hypermutation (Diaz *et al.*, 2001; Zan *et al.*, 2001). As shown by single-cell PCR analysis, *in vivo* gene ablation of Rev3 in mature B cells was also found to reduce the frequency of somatic mutations and leave the pattern of SHM unaffected. As G/C transitions do not depend on Polζ, the phenotype is likely to be caused by the enormous sensitivity of cells to Rev3 ablation. Alternatively, other extender polymerases can take over and the lethality is caused by a TLS-independent role of polymerase ζ.

6.7.6 *Polθ is dispensable during SHM*

Based on its sequence homology, Polθ belongs to the A-family of polymerases (Harris *et al.*, 1996). Polθ lacks exonuclease activity and hence has a relatively high misincorporation frequency of ~10^{-2} to 10^{-3} (Seki *et al.*, 2004). In contrast to the other TLS polymerases, Polθ does not require an extender polymerase. It does not only incorporate nucleotides opposite of abasic sites but can also extend from the inserted nucleotides, which is a unique property.

Both the laboratories of Casali and O-Wang reported on an important role of Polθ in SHM. While the laboratory of Casali reported a dramatic decrease in the frequency of mutations and an increase in G/C transitions, the laboratory of O-Wang reported that Polθ-deficient mice had only a mild reduction in the number of mutations and an increase in G to C transversions. In addition, the O-Wang group analyzed SHM in mice expressing a catalytically-inactive Polθ and found an actual decrease in mutations at template G/C. Given these striking differences, the Gearhart group recently reexamined this issue in Polθ-deficient mice and Polθ/Polη double-deficient mice. Based on the frequency and spectra of the mutations they observed, Polθ has no major role in somatic hypermutation (Martomo *et al.*, 2008; Masuda *et al.*, 2006; Masuda *et al.*, 2005; Zan *et al.*, 2005).

6.7.7 *Other TLS polymerases: Polλ and Polμ*

DNA polymerases Polλ and Polμ have shown robust translesion activity *in vitro*. However, Polλ and Polμ appear dispensable for SHM (Bertocci *et al.*, 2002; Lucas *et al.*, 2005).

In summary, despite their large number, most non-replicative polymerases (except Polν) have been tested for their role in SHM. So far only Polη and Rev1 appear to have non-redundant functions in establishing most A/T and G to C transversions, respectively. Other TLS polymerases are likely to be involved in and might compensate for the absence of Polη and Rev1. The diversity of structurally-related TLS polymerases raises a central question: what regulates the activation of TLS polymerases and inactivation of replicative polymerases during SHM?

6.8 Regulating TLS by Ubiquitylation of PCNA

Non-instructive DNA lesions such as abasic sites, cause problems for high-fidelity polymerases and lead to replication fork stalling. If the "stalling" lesion is not repaired, the replication fork may collapse (Tercero and Diffley, 2001). Such a collapse can generate double-strand breaks, which can in turn trigger cell death (McGlynn and Lloyd, 2002). To prevent such fatal lesions during replication, eukaryotic cells are equipped with two alternative DNA damage tolerance pathways: template switching (damage avoidance) and TLS (damage bypass; Friedberg, 2005; Haracska *et al.*, 2001; Lawrence, 1994; Murli and Walker, 1993). Damage tolerance allows a cell to continue DNA synthesis without an *a priori* repair of the initial lesion. Template switching uses intact DNA of the sister chromatid as a template to continue replication and is therefore error-free (Zhang and Lawrence, 2005). While template switching bypasses the lesion indirectly, TLS enables direct replication across the damaged template. However as previously described, depending on the type of damage and the nature of the TLS polymerase involved, TLS can be highly error-prone.

This raises three central questions: what determines the decision making between conventional repair and damage tolerance; how are low-fidelity polymerases selected/activated; and how does the system decide between error-prone TLS and error-free template switching? Studies in *S. cerevisiae* suggested a mechanism underlying the selective activation of these pathways. Both modes of lesion bypass appear to be controlled by specific posttranslational modifications of the homotrimeric DNA sliding clamp PCNA (Hoege *et al.*, 2002; Fig. 6.1). PCNA tethers DNA polymerases to their substrate and thereby serves as a critical processivity factor for DNA synthesis. The use of PCNA as a sliding clamp for TLS polymerases during damage bypass implies a polymerase switch from the high-fidelity Polδ to low-fidelity TLS polymerases (Plosky and Woodgate, 2004). During replication, Polδ binds PCNA through its PIP (PCNA-interacting peptide) box (Warbrick, 1998). At this stage TLS polymerases associate weakly with PCNA. When the high-fidelity replication machinery is stalled upon encountering a lesion, PCNA becomes monoubiquitylated at its lysine residue 164 ($PCNA^{K164}$; Hoege *et al.*, 2002). At that moment TLS polymerases are recruited to the monoubiquitylated PCNA (PCNA-Ub) through the combined affinity of the PIP box and ubiquitin-binding domains, i.e. a Ub-binding motif (UBM) or a Ub-binding zinc finger (UBZ) resulting in a transient displacement of the high-fidelity polymerase Polδ (Bienko *et al.*, 2005). The ubiquitin-conjugating/ligating complex Rad6/Rad18 (E2/E3) mediates site-specific monoubiquitylation of PCNA and thereby is thought to enable polymerase switching and activation of TLS-dependent damage tolerance. The alternative pathway of damage tolerance, template switching, requires further polyubiquitylation of the monoubiquitylated PCNA (Hoege *et al.*, 2002). The heterodimeric E2 ubiquitin conjugase consisting of Ubc13 and Mms2 cooperates with the RING finger E3 ligase Rad5 in specific lysine 63-linked polyubiquitylation of PCNA-Ub (Torres-Ramos *et al.*, 2002). How polyubiquitylated PCNA mechanistically activates the error-free branch of DNA damage tolerance and the relevance of this pathway in mammals is the focus of current research.

The fact that the RAD6 epistasis group has functional orthologs in higher eukaryotes, suggested that this pathway is of general importance. In support of this notion, UV-irradiation of both human and murine cells was shown to lead to the monoubiquitylation of PCNA at the conserved K164 residue, which resulted in the accumulation of TLS polymerases at sites of DNA damage (Kannouche *et al.*, 2004). This implies a conserved mechanism between yeast and mammals in the recruitment and activation of TLS polymerases.

6.9 SHM: Mutagenesis at Template A/T Requires PCNA-Ub

To test whether this mode of polymerase activation and inactivation is operative in mammalian cells and related to the generation of somatic mutations in hypermutating B cells, PCNA mutant mice that contain a lysine 164 to arginine mutation ($PCNA^{K164R}$) have been generated. This subtle point mutation prohibits $PCNA^{K164}$ modifications without interfering with other pivotal functions of the protein. Analysis of the mutation spectrum of mutated Ig genes in B cells from these knock-in mice revealed a 10-fold reduction in A/T mutations (Langerak *et al.*, 2007). In agreement, PCNA knock-out mice reconstituted with a $PCNA^{K164R}$ transgene showed a reduction of A/T mutations in Ig genes (Roa *et al.*, 2008). Interestingly, the combined failure in removing uracils by UNG2 and the ubiquination of PCNA results in a mutation spectrum in which almost all mutations (99.3%) are G/C transition mutations (Krijger *et al.*, 2009), a finding comparable to Ung/MutSα double mutants previously described (Rada *et al.*, 2004; Shen *et al.*, 2006). These data strongly indicate that most A/T mutations introduced downstream of MutSα are regulated by PCNA-Ub. As A/T mutations depend mainly on polymerase η (and, in its absence, polymerase κ and at least a third yet unidentified polymerase [Faili *et al.*, 2009]), these data indicate that during MSH2-dependent SHM both Polη and Polκ depend on PCNA-Ub to establish most A/T mutations. Furthermore, UNG2-dependent A/T mutations, which have recently been shown to be introduced by polymerase η (Delbos *et al.*, 2007), disappear almost completely in $PCNA^{K164R}$/MSH2 mutant B cells, indicating that most Ung-dependent A/T mutations require PCNA-Ub to activate Polη. In summary, these data suggest a model in which PCNA-Ub acts downstream of both MSH2 and UNG2 to ensure that TLS Polη (and, in its absence, also Polκ) are recruited to introduce mutations at template A/T. The remaining A/T mutator activity found in somatically-mutated Ig genes of $PCNA^{K164R}$ mutant B cells might be generated in a PCNA-Ub-independent manner by Polη, Polκ or a yet unidentified polymerase.

6.10 PCNA-Ub-Independent G/C Transversions During SHM

Surprisingly, while most A/T mutations depend on PCNA-Ub, the generation of G/C transversions is unaltered in $PCNA^{K164R}$ mutant GC B cells. Given the role of the TLS polymerase Rev1 in generating G to C transversions during SHM (Jansen *et al.*, 2006; Ross and Sale, 2006), these findings exclude a role for PCNA-Ub in activating Rev1 and all other yet unidentified 'G/C transverters' during SHM in mammals. In agreement, damage tolerance mediated by Rev1 was found to be independent of PCNA-Ub in the chicken DT40 B cell line (Edmunds *et al.*, 2008). In contrast, in DT40 cells Rev1 depends on PCNA-Ub to generate G to C transversions during SHM (Arakawa *et al.*, 2006). This raises the question of how are G to C transversions controlled during SHM in mammals?

The heterotrimeric Rad9/Rad1/Hus1 complex (also known as 9-1-1 complex) is structurally very similar to PCNA (Dore *et al.*, 2009).

As shown in yeast, 9-1-1 interacts with TLS polymerases and induces TLS independent of PCNA^{K164} modification. Interestingly, the 9-1-1 complex has been identified as an alternative substrate of the Rad6/18 ubiquitin conjugase/ligase complex (Fu *et al.*, 2008). The 9-1-1 complex may therefore provide a platform for PCNA-independent TLS during SHM and a closer analysis of this sliding clamp may reveal a regulatory role in SHM.

In summary, monoubiquitylation of PCNA^{K164} and activation of the TLS polymerase has a dual-mutagenic and anti-mutagenic physiological purpose. During SHM, Polη is recruited by PCNA-Ub to generate almost all A/T mutations. During replication across UV-induced lesions, Polη is recruited to correctly insert AA opposite a *cis*-syn TT-CPD to increase the fitness of both yeast and higher eukaryotes in response to genotoxic stress. Given the defined mutation signatures of individual TLS polymerases, the analysis of SHM provides an effective read-out system for the activity of diverse TLS polymerases and their regulation in mammals. While ubiquitylation of PCNA serves to generate A/T mutations downstream of MutSα, G/C transversions may depend on 9-1-1.

6.11 MutSα and UNG2 do not Compete During SHM: Cell Cycle and Error-Free Repair

Whether the UNG2 and MSH2 pathways of SHM compete for the recognition of a U or act non-competitively is a long-standing question (Di Noia and Neuberger, 2007; Weill and Reynaud, 2008). In the latter case, the common substrate U is expected to be removed to exclude that a U not processed by UNG2 enters MSH2-dependent mutagenesis and *vice versa.* Recently, this issue was addressed by comparing the absolute efficacies in generating defined nucleotide substitutions between WT, Ung and MSH2-deficient mouse models. If MSH2-dependent mutagenesis would compete with UNG2 mutagenesis, one expects A/T mutations to increase in the absence of UNG2. However, as the efficacy in generating A/T mutations does not change in Ung-deficient mice it was concluded that MSH2-dependent SHM does not compete with

UNG2-dependent mutagenesis, a finding consistent with the non-competitive model of SHM. In line with this model, MSH2 seems incapable of removing uracils usually processed by UNG2, as revealed by the selective and compensatory increase of G/C transitions as a consequence of direct replication over the U. In summary, these data indicate that uracils normally recognized and processed by UNG2 remain refractory to MSH2-dependent SHM in UNG2-deficient B cells.

The reverse situation, whether UNG2 is capable of removing uracils normally processed by MSH2, was addressed by comparing the efficacy in the generation of G/C transversions between WT and MSH2-deficient B cells. If UNG2 mutagenesis would compete with MSH2, one expects G/C transversions to increase in the absence of MSH2. However, as mentioned previously, the lack of mismatch recognition results in a two-fold decrease in the frequency of these mutations. Furthermore, the lack of mismatch recognition does not change the efficacy in generating G/C transitions, suggesting that prior to replication, U usually processed by MSH2 are now repaired by BER. These data further support the non-competitive model. A model of how the separation of MSH2- and UNG2-dependent mutation can be achieved was proposed recently by J.C. Weill and C.A. Reynaud (Weill and Reynaud, 2008). In this model uracils which may arise as a consequence of processive AID activity (Storb *et al.*, 2009; Pham *et al.*, 2003) are introduced on both strands at distinct phases of the cell cycle (Aoufouchi *et al.*, 2008). U introduced during the S phase can directly act as template T during replication to generate G/C transitions or be converted to an AP site by UNG2. If not repaired, the AP site causes replicative polymerase to stall and activate a TLS polymerase to initiate the generation of G/C transversions (and possibly transitions) and some A/T mutations. Uracils generated outside the S-phase would be recognized as a U:G mismatch by MSH2-MSH6, resulting in mutations at template A/T. In support of this model, UNG is strongly upregulated in the transition from the G1 to S phase (Hagen *et al.*, 2008). In, addition, Polη is recruited upon DNA damage in cells arrested in G1 (Soria *et al.*, 2009). Although MMR normally is post-replicative, components such as MutSα and Exo1 may function outside of the S phase to support SHM.

6.12 Aberrant Targeting of AID and Error-Free Repair of AID-Induced Uracils

It has been known for over a decade, that non-Ig genes can be mutated during SHM. Memory B cells isolated from human peripheral blood were found to have mutations in the BCL6 gene (Shen *et al.*, 1998). The BCL6 gene was sequenced in the first intron from 0.64 to 1.43kb from the start of transcription and the peak of mutations was found in the first half of this region. No mutations were seen further 3′ between 2.4 to 3.2kb from the promoter. The mutation frequencies in the hypermutable region were 2×10^{-3} to 7×10^{-4}. In one donor, where the mutation frequency of BCL6 was 2×10^{-3}, IgH genes had mutations at a 5×10^{-2} frequency, in the normal range of IgH mutations in humans. Thus, BCL6 mutations are at a one to two orders lower frequency than Ig gene mutations, but the pattern of mutations is the same as that of Ig genes. No BCL6 mutations were seen in resting B cells from the same donors. These findings were confirmed in tonsillar germinal center B cells and B cell lymphomas (Pasqualucci *et al.*, 1998; Peng *et al.*, 1999).

Mutations were also found in other non-Ig genes. About 15% of germinal center and memory, but not naïve B cells of normal human donors, were found to have mutations in the CD95 gene (Muschen *et al.*, 2000). Also, the Ig-alpha and Ig-beta genes were found mutated in malignant and normal human B cells (Gordon *et al.*, 2003). Mutations were also found in four proto-oncogenes in diffuse large-cell lymphomas that express AID: PIM1, MYC, RhoH/TTF and PAX5 (Pasqualucci *et al.*, 2001). Initially, these genes appeared not mutated in normal human germinal center B cells or in other germinal-center-derived lymphomas. However, an extended analysis of these genes in germinal center B cells of wild-type mice revealed that these highly expressed genes do mutate in an AID-dependent manner, albeit at a low frequency (Liu *et al.*, 2008). It appeared thus, that these non-Ig genes are targeted by AID and may have *cis*-elements that can attract AID. Furthermore, these data indicate that certain germinal center-derived lymphomas are prone to accumulate mutations in non-Ig genes.

Other non-Ig genes were found not to be mutated significantly in normal human memory B cells: c-MYC was mutated at the very low frequency of ~1 x10^{-4}: no mutations were found in ribosomal small subunit protein S14, α-fetoprotein, TBP and survivin (Shen *et al.*, 1998; Shen *et al.*, 2000). However, many of the expressed non-Ig genes were mutated in germinal center B cells of Msh2/Ung double-deficient mice (Liu *et al.*, 2008). The same genes were mutated at 0.6 to 17.4 lower frequencies in wild-type mice. The list included the c-Myc gene found in human memory B cells to be mutated at a frequency of ~1 x10^{-4} (Shen *et al.*, 1998). Myc was mutated in the double-knockout mice at about a 10x higher frequency of ~1 x10^{-3} (Liu *et al.*, 2008). It was concluded that these non-Ig genes suffered C to U deaminations by AID but that the uracils were repaired error-free and the authors considered that there were different mechanisms of dealing with AID-induced uracils: error-prone in Ig and BCL6 genes, but error-free in other genes.

This idea assumes that uracils in Ig genes are not repaired in an error-free fashion. However, a recent analysis of MMR/BER double-deficient mice showed that AID-induced uracils are repaired efficiently in Ig genes (Storb *et al.*, 2009). The data indicated that 39 to 47% of the mutations in Ig genes of the double-knockout mice were repaired error-free in wild-type mice. This percentage is likely to be an underestimate, as the number of G/C transitions found in Ung/MMR-/- mice only arose from uracils that were not processed by Ung; additionally, other redundant uracil glycosylases such as Smug1 (Di Noia *et al.*, 2006) – see previous sections in this chapter – may compete for repair of such lesions. The actual number of G/C mutations is likely to be higher if the other uracil glycosylases were also absent. Therefore, the proportion of uracils repaired error-free (39–47%) in Ig genes is likely to be an underestimate.

Assuming that error-free repair plays an equally major role in curtailing mutations in Ig genes as well as non-Ig genes, one wonders whether or not there is a special mechanism active during SHM to submit any gene to error-prone repair. Most of the error-free repair of uracils in Ig genes is apparently due to Ung rather than MMR (Storb *et al.*, 2009). Possibly, there is a random chance for error-free repair occurring before error-prone mechanisms can act, as discussed in other sections of this chapter. Ung is also involved in error-prone repair, since in MMR-deficient mice, transversions at G/C and mutations at A/T still occur (Rada *et al.*, 1998; Frey *et al.*, 1998). Thus, it is possible that Ung's activity is neutral in outcome. Clearly, Ig genes are targeted at much higher frequencies than any non-Ig genes. This is presumably due to the combination of high transcription rates and a high frequency and intragenic location of a *cis*-acting element, CAGGTG, which has been shown to be sufficient to attract AID to a target gene (Tanaka *et al.*, 2010). With lower transcription rates and fewer CAGGTG elements, non-Ig genes acquire many fewer AID-induced uracils than Ig genes. Perhaps the uracil load is generally low enough in non-Ig genes to ensure that most are repaired error-free. Ig genes, on the other hand, may become overloaded with uracils, so that abasic sites created by Ung (and perhaps other uracil glycosylases: see previous sections of this chapter) lead to the recruitment of lesion bypass, error-prone polymerases and, therefore, stable mutations. This scenario has some support in the finding that over-expression of AID causes SHM in various genes and mutations even in non-germinal center B cells and non-B cells. AID transfected into B cell hybridomas caused mutation of the endogenous VH genes (Martin *et al.*, 2002). Transgenic AID in Chinese hamster ovary cells caused mutations in the AID transgene itself (Martin and Scharff, 2002). Transgenic AID also caused mutations in an artificial GFP substrate in NIH 3T3 fibroblasts (Yoshikawa *et al.*, 2002). Thus, AID does not appear to require germinal center B cell-specific co-factors to cause mutations when it is over-expressed. However, the marked shift of the mutation signature towards G/C transition mutations suggests that these mutations do not require error-prone repair and simply arise as a consequence of replication across U. The balance between error-free and error-prone repair, as well as the signature of SHM is likely to depend on several factors, i.e. the number of cytidine deaminations, the cell-cycle stage when lesions are introduced, the sequence context, the (in)activity of DNA repair, DNA damage tolerance and DNA damage response pathways.

6.13 Acknowledgements

The work described in this chapter was supported by The Netherlands Organisation for Scientific Research (VIDI program NWO 917.56.328 to H.J.), SFN (SFN 2.129 to H.J) the Dutch Cancer Society (KWF project NKI 2008-4112 to H.J.) and NIH grants (AI047380 and AI081167 to U.S.).

6.14 References

1. Aoufouchi S., Faili A., Zober C. *et al.* (2008). Proteasomal degradation restricts the nuclear lifespan of AID. *J Exp Med* **205**: 1357–1368.
2. Arakawa H., Moldovan G.L., Saribasak H. *et al.* (2006). A role for PCNA ubiquitination in immunoglobulin hypermutation. *PLoS Biol* **4**: e366.
3. Bardwell P.D., Martin A., Wong E. *et al.* (2003). Cutting edge: the G-U mismatch glycosylase methyl-CpG binding domain 4 is dispensable for somatic hypermutation and class switch recombination. *J Immunol* **170**: 1620–1624.
4. Bardwell P.D., Woo C.J., Wei K. *et al.* (2004). Altered somatic hypermutation and reduced class-switch recombination in exonuclease 1-mutant mice. *Nat Immunol* **5**: 224–229.
5. Bertocci B., De Smet A., Flatter E. *et al.* (2002). Cutting edge: DNA polymerases mu and lambda are dispensable for Ig gene hypermutation. *J Immunol* **168**: 3702–3706.
6. Bienko M., Green C.M., Crosetto N. *et al.* (2005). Ubiquitin-binding domains in Y-family polymerases regulate translesion synthesis. *Science* **310**: 1821–1824.
7. Brenner S., Milstein C. (1966). Origin of antibody variation. *Nature* **211**: 242–243.
8. Chaudhuri J., Tian M., Khuong C. *et al.* (2003). Transcription-targeted DNA deamination by the AID antibody diversification enzyme. *Nature* **422**: 726–730.
9. Delbos F., Aoufouchi S., Faili A. *et al.* (2007). DNA polymerase eta is the sole contributor of A/T modifications during immunoglobulin gene hypermutation in the mouse. *J Exp Med* **204**: 17–23.
10. Delbos F., De Smet A., Faili A. *et al.* (2005). Contribution of DNA polymerase eta to immunoglobulin gene hypermutation in the mouse. *J Exp Med* **201**: 1191–1196.
11. Di Noia J.M., Neuberger M.S. (2007). Molecular mechanisms of antibody somatic hypermutation. *Annu Rev Biochem* **76**: 1–22.
12. Di Noia J.M., Rada C., Neuberger M.S. (2006). SMUG1 is able to excise uracil from immunoglobulin genes: insight into mutation versus repair. *EMBO J* **25**: 585–595.
13. Diaz M., Verkoczy L.K., Flajnik M.F. *et al.* (2001). Decreased frequency of somatic hypermutation and impaired affinity maturation but intact germinal center formation in mice expressing antisense RNA to DNA polymerase zeta. *J Immunol* **167**: 327–335.

14. Dickerson S.K., Market E., Besmer E. *et al.* (2003). AID mediates hypermutation by deaminating single stranded DNA. *J Exp Med* **197**: 1291–1296.
15. Dore A.S., Kilkenny M.L., Rzechorzek N.J. *et al.* (2009). Crystal structure of the rad9-rad1-hus1 DNA damage checkpoint complex – implications for clamp loading and regulation. *Mol Cell* **34**: 735–745.
16. Dorner T., Foster S.J., Farner N.L. *et al.* (1998). Somatic hypermutation of human immunoglobulin heavy chain genes: targeting of RGYW motifs on both DNA strands. *Eur J Immunol* **28**: 3384–3396.
17. Duncan B.K., Miller J.H. (1980). Mutagenic deamination of cytosine residues in DNA. *Nature* **287**: 560–561.
18. Edmunds C.E., Simpson L.J., Sale J.E. (2008). PCNA ubiquitination and REV1 define temporally distinct mechanisms for controlling translesion synthesis in the avian cell line DT40. *Mol Cell* **30**: 519–529.
19. Ehrenstein M.R., Rada C., Jones A.M. *et al.* (2001). Switch junction sequences in PMS2-deficient mice reveal a microhomology-mediated mechanism of Ig class switch recombination. *Proc Natl Acad Sci USA* **98**: 14553–14558.
20. Esposito G., Texido G., Betz U.A. *et al.* (2000). Mice reconstituted with DNA polymerase beta-deficient fetal liver cells are able to mount a T cell-dependent immune response and mutate their Ig genes normally. *Proc Natl Acad Sci USA* **97**: 1166–1171.
21. Faili A., Stary A., Delbos F. *et al.* (2009). A backup role of DNA polymerase kappa in Ig gene hypermutation only takes place in the complete absence of DNA polymerase eta. *J Immunol* **182**: 6353–6359.
22. Frey S., Bertocci B., Delbos F. *et al.* (1998). Mismatch repair deficiency interferes with the accumulation of mutations in chronically stimulated B cells and not with the hypermutation process. *Immunity* **9**: 127–134.
23. Friedberg E.C. (2005). Suffering in silence: the tolerance of DNA damage. *Nat Rev Mol Cell Biol* **6**: 943–953.
24. Friedberg E.C., Walker G.C., Siede W. *et al.* (2006). *DNA repair and mutagenesis*, Washington, D.C., ASM Press.
25. Fu Y., Zhu Y., Zhang K. *et al.* (2008). Rad6-Rad18 mediates a eukaryotic SOS response by ubiquitinating the 9-1-1 checkpoint clamp. *Cell* **133**: 601–611.
26. Gan G.N., Wittschieben J.P., Wittschieben B.O. *et al.* (2008). DNA polymerase zeta (pol zeta) in higher eukaryotes. *Cell Res* **18**: 174–183.
27. Gening L.V., Tarantul V.Z. (2006). Involvement of DNA polymerase iota in hypermutagenesis and tumorigenesis: an improper model can lead to inaccurate results. *Immunol Lett* **106**: 198–199.
28. Gerlach V.L., Aravind L., Gotway G. *et al.* (1999). Human and mouse homologs of Escherichia coli DinB (DNA polymerase IV), members of the UmuC/DinB superfamily. *Proc Natl Acad Sci USA* **96**: 11922–11927.
29. Gordon M.S., Kanegai C.M., Doerr J.R. *et al.* (2003). Somatic hypermutation of the B cell receptor genes B29 (Igbeta, CD79b) and mb1 (Igalpha, CD79a). *Proc Natl Acad Sci USA* **100**: 4126–4131.
30. Hagen L., Kavli B., Sousa M.M. *et al.* (2008). Cell cycle-specific UNG2 phosphorylations regulate protein turnover, activity and association with RPA. *EMBO J* **27**: 51–61.

31. Haracska L., Washington M.T., Prakash S. *et al.* (2001). Inefficient bypass of an abasic site by DNA polymerase eta. *J Biol Chem* **276**: 6861–6866.
32. Harris P.V., Mazina O.M., Leonhardt E.A. *et al.* (1996). Molecular cloning of Drosophila mus308, a gene involved in DNA cross-link repair with homology to prokaryotic DNA polymerase I genes. *Mol Cell Biol* **16**: 5764–5771.
33. Hoege C., Pfander B., Moldovan G.L. *et al.* (2002). RAD6-dependent DNA repair is linked to modification of PCNA by ubiquitin and SUMO. *Nature* **419**: 135–141.
34. Jacobs H., Fukita Y., Van Der Horst G.T. *et al.* (1998). Hypermutation of immunoglobulin genes in memory B cells of DNA repair-deficient mice. *J Exp Med* **187**: 1735–1743.
35. Jansen J.G., Fousteri M.I., De Wind N. (2007). Send in the clamps: control of DNA translesion synthesis in eukaryotes. *Mol Cell* **28**: 522–529.
36. Jansen J.G., Langerak P., Tsaalbi-Shtylik A. *et al.* (2006). Strand-biased defect in C/G transversions in hypermutating immunoglobulin genes in Rev1-deficient mice. *J Exp Med* **203**: 319–323.
37. Jansen J.G., Tsaalbi-Shtylik A., Hendriks G. *et al.* (2009). Mammalian polymerase zeta is essential for post-replication repair of UV-induced DNA lesions. *DNA Repair* **8**: 1444–1451.
38. Jansen J.G., Tsaalbi-Shtylik A., Langerak P. *et al.* (2005). The BRCT domain of mammalian Rev1 is involved in regulating DNA translesion synthesis. *Nucleic Acids Res* **33**: 356–365.
39. Jiricny J. (1998). Replication errors: cha(lle)nging the genome. *EMBO J* **17**: 6427–6436.
40. Johnson R.E., Prakash S., Prakash L. (1999). Efficient bypass of a thymine-thymine dimer by yeast DNA polymerase, Poleta. *Science* **283**: 1001–1004.
41. Johnson R.E., Washington M.T., Haracska L. *et al.* (2000). Eukaryotic polymerases iota and zeta act sequentially to bypass DNA lesions. *Nature* **406**: 1015–1019.
42. Kadyrov F.A., Dzantiev L., Constantin N. *et al.* (2006). Endonucleolytic function of MutLalpha in human mismatch repair. *Cell* **126**: 297–308.
43. Kannouche P.L., Wing J., Lehmann A.R. (2004). Interaction of human DNA polymerase eta with monoubiquitinated PCNA: a possible mechanism for the polymerase switch in response to DNA damage. *Mol Cell* **14**: 491–500.
44. Krijger P.H., Langerak P., Van Den Berk P.C. *et al.* (2009). Dependence of nucleotide substitutions on Ung2, Msh2, and PCNA-Ub during somatic hypermutation. *J Exp Med* **206**: 2603–2611.
45. Krokan H.E., Drablos F., Slupphaug G. (2002). Uracil in DNA – occurrence, consequences and repair. *Oncogene* **21**: 8935–8948.
46. Kunkel T.A. (1985). The mutational specificity of DNA polymerase-beta during in vitro DNA synthesis. Production of frameshift, base substitution, and deletion mutations. *J Biol Chem* **260**: 5787–5796.
47. Langerak P., Nygren A.O., Krijger P.H. *et al.* (2007). A/T mutagenesis in hypermutated immunoglobulin genes strongly depends on PCNAK164 modification. *J Exp Med* **204**: 1989–1998.

48. Larsen E., Kleppa L., Meza T.J. *et al.* (2008). Early-onset lymphoma and extensive embryonic apoptosis in two domain-specific Fen1 mice mutants. *Cancer Res* **68**: 4571–4579.
49. Larson E.D., Cummings W.J., Bednarski D.W. *et al.* (2005). MRE11/RAD50 cleaves DNA in the AID/UNG-dependent pathway of immunoglobulin gene diversification. *Mol Cell* **20**: 367–375.
50. Lawrence C. (1994). The RAD6 DNA repair pathway in Saccharomyces cerevisiae: what does it do, and how does it do it? *Bioessays* **16**: 253–258.
51. Liu M., Duke J.L., Richter D.J. *et al.* (2008). Two levels of protection for the B cell genome during somatic hypermutation. *Nature* **451**: 841–845.
52. Lucas D., Lain De Lera T., Gonzalez M.A. *et al.* (2005). Polymerase mu is up-regulated during the T cell-dependent immune response and its deficiency alters developmental dynamics of spleen centroblasts. *Eur J Immunol* **35**: 1601–1611.
53. Martin A., Bardwell P.D., Woo C.J. *et al.* (2002). Activation-induced cytidine deaminase turns on somatic hypermutation in hybridomas. *Nature* **415**: 802–806.
54. Martin A., Scharff M.D. (2002). Somatic hypermutation of the AID transgene in B and non-B cells. *Proc Natl Acad Sci. USA* **99**: 12304–12308.
55. Martomo S.A., Saribasak H., Yokoi M. *et al.* (2008). Reevaluation of the role of DNA polymerase theta in somatic hypermutation of immunoglobulin genes. *DNA Repair* **7**: 1603–1608.
56. Martomo S.A., Yang W.W., Gearhart P.J. (2004). A role for Msh6 but not Msh3 in somatic hypermutation and class switch recombination. *J Exp Med* **200**: 61–68.
57. Martomo S.A., Yang W.W., Wersto R.P. *et al.* (2005). Different mutation signatures in DNA polymerase eta- and MSH6-deficient mice suggest separate roles in antibody diversification. *Proc Natl Acad Sci USA* **102**: 8656–8661.
58. Masuda K., Ouchida R., Hikida M. *et al.* (2006). Absence of DNA polymerase theta results in decreased somatic hypermutation frequency and altered mutation patterns in Ig genes. *DNA Repair* **5**: 1384–1391.
59. Masuda K., Ouchida R., Takeuchi A. *et al.* (2005). DNA polymerase theta contributes to the generation of C/G mutations during somatic hypermutation of Ig genes. *Proc Natl Acad Sci USA* **102**: 13986–13991.
60. Masutani C., Kusumoto R., Yamada A. *et al.* (1999). The XPV (xeroderma pigmentosum variant) gene encodes human DNA polymerase eta. *Nature* **399**: 700–704.
61. McDonald J.P., Frank E.G., Plosky B.S. *et al.* (2003). 129-derived strains of mice are deficient in DNA polymerase iota and have normal immunoglobulin hypermutation. *J Exp Med* **198**: 635–643.
62. McGlynn P., Lloyd R.G. (2002). Replicating past lesions in DNA. *Mol Cell* **10**: 700–701.
63. Milstein C., Neuberger M.S., Staden R. (1998). Both DNA strands of antibody genes are hypermutation targets. *Proc Natl Acad Sci USA* **95**: 8791–8794.
64. Modrich P. (2006). Mechanisms in eukaryotic mismatch repair. *J Biol Chem* **281**: 30305–30309.
65. Muramatsu M., Kinoshita K., Fagarasan S. *et al.* (2000). Class switch recombination and hypermutation require activation-induced cytidine deaminase (AID), a potential RNA editing enzyme. *Cell* **102**: 553–563.

66. Murli S., Walker G.C. (1993). SOS mutagenesis. *Curr Opin Genet Dev* **3**: 719–725.
67. Muschen M., Re D., Jungnickel B., Diehl V. *et al.* (2000). Somatic mutation of the CD95 gene in human B cells as a side-effect of the germinal center reaction. *J Exp Med* **192**: 1833–1840.
68. Nelson J.R., Lawrence C.W., Hinkle D.C. (1996). Deoxycytidyl transferase activity of yeast REV1 protein. *Nature* **382**: 729–731.
69. Ohashi E., Bebenek K., Matsuda T. *et al.* (2000). Fidelity and processivity of DNA synthesis by DNA polymerase kappa, the product of the human DINB1 gene. *J Biol Chem* **275**: 39678–39684.
70. Pasqualucci L., Migliazza A., Fracchiolla N. *et al.* (1998). BCL-6 mutations in normal germinal center B cells: evidence of somatic hypermutation acting outside Ig loci. *Proc Natl Acad Sci USA* **95**: 11816–11821.
71. Pasqualucci L., Neumeister P., Goossens T. *et al.* (2001). Hypermutation of multiple proto-oncogenes in B-cell diffuse large-cell lymphomas. *Nature* **412**: 341–346.
72. Peng H.Z., Du M.Q., Koulis A. *et al.* (1999). Nonimmunoglobulin gene hypermutation in germinal center B cells. *Blood* **93**: 2167–2172.
73. Pham P., Bransteitter R., Petruska J. *et al.* Processive AID-catalysed cytosine deamination on single-stranded DNA simulates somatic hypermutation. *Nature* **424**: 103–107.
74. Phung Q.H., Winter D.B., Alrefai R. *et al.* (1999). Hypermutation in Ig V genes from mice deficient in the MLH1 mismatch repair protein. *J Immunol* **162**: 3121–3124.
75. Phung Q.H., Winter D.B., Cranston A. *et al.* (1998). Increased hypermutation at G and C nucleotides in immunoglobulin variable genes from mice deficient in the MSH2 mismatch repair protein. *J Exp Med* **187**: 1745–1751.
76. Plosky B.S., Woodgate R. (2004). Switching from high-fidelity replicases to low-fidelity lesion-bypass polymerases. *Curr Opin Genet Dev* **14**: 113–119.
77. Podlutsky A.J., Dianova, Ii, Podust V.N. *et al.* (2001). Human DNA polymerase beta initiates DNA synthesis during long-patch repair of reduced AP sites in DNA. *EMBO J* **20**: 1477–1482.
78. Prakash S., Johnson R.E., Prakash L. (2005). Eukaryotic translesion synthesis DNA polymerases: specificity of structure and function. *Annu Rev Biochem* **74**: 317–353.
79. Rada C., Di Noia J.M., Neuberger M.S. (2004). Mismatch recognition and uracil excision provide complementary paths to both Ig switching and the A/T-focused phase of somatic mutation. *Mol Cell* **16**: 163–171.
80. Rada C., Ehrenstein M.R., Neuberger M.S. *et al.* (1998). Hot spot focusing of somatic hypermutation in MSH2-deficient mice suggests two stages of mutational targeting. *Immunity* **9**: 135–141.
81. Rada C., Williams G.T., Nilsen H. *et al.* (2002). Immunoglobulin isotype switching is inhibited and somatic hypermutation perturbed in UNG-deficient mice. *Curr Biol* **12**: 1748–1755.

82. Ramiro A.R., Stavropoulos P., Jankovic M. *et al.* (2003). Transcription enhances AID-mediated cytidine deamination by exposing single-stranded DNA on the nontemplate strand. *Nat Immunol* **4**: 452–456.
83. Roa S., Avdievich E., Peled J.U. *et al.* (2008). Ubiquitylated PCNA plays a role in somatic hypermutation and class-switch recombination and is required for meiotic progression. *Proc Natl Acad Sci USA* **105**: 16248–16253.
84. Rogozin I.B., Pavlov Y.I., Bebenek K. *et al.* (2001). Somatic mutation hotspots correlate with DNA polymerase eta error spectrum. *Nat Immunol* **2**: 530–536.
85. Ross A.L., Sale J.E. (2006). The catalytic activity of REV1 is employed during immunoglobulin gene diversification in DT40. *Mol Immunol* **43**: 1587–1594.
86. Schanz S., Castor D., Fischer F. *et al.* (2009). Interference of mismatch and base excision repair during the processing of adjacent U/G mispairs may play a key role in somatic hypermutation. *Proc Natl Acad Sci USA* **106**: 5593–5598.
87. Schenten D., Gerlach V.L., Guo C. *et al.* (2002). DNA polymerase kappa deficiency does not affect somatic hypermutation in mice. *Eur J Immunol* **32**: 3152–3160.
88. Seki M., Masutani C., Yang L.W. *et al.* (2004). High-efficiency bypass of DNA damage by human DNA polymerase Q. *EMBO J* **23**: 4484–4494.
89. Shachar S., Ziv O., Avkin S. *et al.* (2009). Two-polymerase mechanisms dictate error-free and error-prone translesion DNA synthesis in mammals. *EMBO J* **28**: 383–393.
90. Shen H.M., Michael N., Kim N. *et al.* (2000). The TATA binding protein, c-Myc and survivin genes are not somatically hypermutated, while Ig and BCL6 genes are hypermutated in human memory B cells. *Int Immunol* **12**: 1085–1093.
91. Shen H.M., Peters A., Baron B. *et al.* (1998). Mutation of BCL-6 gene in normal B cells by the process of somatic hypermutation of Ig genes. *Science* **280**: 1750–1752.
92. Shen H.M., Tanaka A., Bozek G. *et al.* (2006). Somatic hypermutation and class switch recombination in Msh6(-/-)Ung(-/-) double-knockout mice. *J Immunol* **177**: 5386–5392.
93. Shimizu T., Shinkai Y., Ogi T. *et al.* (2003). The absence of DNA polymerase kappa does not affect somatic hypermutation of the mouse immunoglobulin heavy chain gene. *Immunol Lett* **86**: 265–270.
94. Singhal R.K., Wilson S.H. (1993). Short gap-filling synthesis by DNA polymerase beta is processive. *J Biol Chem* **268**: 15906–15911.
95. Sonoda E., Okada T., Zhao G.Y. *et al.* (2003). Multiple roles of Rev3, the catalytic subunit of polzeta in maintaining genome stability in vertebrates. *EMBO J* **22**: 3188–3197.
96. Soria G., Belluscio L., Van Cappellen W.A. *et al.* (2009). DNA damage induced Pol eta recruitment takes place independently of the cell cycle phase. *Cell Cycle* **8**: 3340–3348.
97. Storb U., Shen H.M., Nicolae D. (2009). Somatic hypermutation: processivity of the cytosine deaminase AID and error-free repair of the resulting uracils. *Cell Cycle* **8**: 3097–3101.
98. Tanaka A., Shen H.M., Ratnam S. *et al.* (2010). Attracting AID to targets of somatic hypermutation. *J Exp Med* **207**: 405–415.

99. Tercero J.A., Diffley J.F. (2001). Regulation of DNA replication fork progression through damaged DNA by the Mec1/Rad53 checkpoint. *Nature* **412**: 553–557.
100. Tissier A., McDonald J.P., Frank E.G. *et al.* (2000). poliota, a remarkably error-prone human DNA polymerase. *Genes Dev* **14**: 1642–1650.
101. Torres-Ramos C.A., Prakash S., Prakash L. (2002). Requirement of RAD5 and MMS2 for postreplication repair of UV-damaged DNA in Saccharomyces cerevisiae. *Mol Cell Biol* **22**: 2419–2426.
102. Warbrick E. (1998). PCNA binding through a conserved motif. *Bioessays* **20**: 195–199.
103. Washington M.T., Johnson R.E., Prakash L. *et al.* (2002). Human DINB1-encoded DNA polymerase kappa is a promiscuous extender of mispaired primer termini. *Proc Natl Acad Sci USA* **99**: 1910–1914.
104. Weill J.C., Reynaud C.A. (2008). DNA polymerases in adaptive immunity. *Nat Rev Immunol* **8**: 302–312.
105. Wiesendanger M., Kneitz B., Edelmann W. *et al.* (2000). Somatic hypermutation in MutS homologue (MSH)3-, MSH6-, and MSH3/MSH6-deficient mice reveals a role for the MSH2-MSH6 heterodimer in modulating the base substitution pattern. *J Exp Med* **191**: 579–584.
106. Wilson T.M., Vaisman A., Martomo S.A. *et al.* (2005). MSH2-MSH6 stimulates DNA polymerase eta, suggesting a role for A:T mutations in antibody genes. *J Exp Med* **201**: 637–645.
107. Winter D.B., Phung Q.H., Umar A. *et al.* (1998). Altered spectra of hypermutation in antibodies from mice deficient for the DNA mismatch repair protein PMS2. *Proc Natl Acad Sci USA* **95**: 6953–6958.
108. Xanthoudakis S., Smeyne R.J., Wallace J.D. *et al.* (1996). The redox/DNA repair protein, Ref-1, is essential for early embryonic development in mice. *Proc Natl Acad Sci USA* **93**: 8919–8923.
109. Yabuki M., Fujii M.M., Maizels N. (2005). The MRE11-RAD50-NBS1 complex accelerates somatic hypermutation and gene conversion of immunoglobulin variable regions. *Nat Immunol* **6**: 730–736.
110. Yoshikawa K., Okazaki I.M., Eto T. *et al.* (2002). AID enzyme-induced hypermutation in an actively transcribed gene in fibroblasts. *Science* **296**: 2033–2036.
111. Zan H., Komori A., Li Z. *et al.* (2001). The translesion DNA polymerase zeta plays a major role in Ig and bcl-6 somatic hypermutation. *Immunity* **14**: 643–653.
112. Zan H., Shima N., Xu Z. *et al.* (2005). The translesion DNA polymerase theta plays a dominant role in immunoglobulin gene somatic hypermutation. *EMBO J* **24**: 3757–3769.
113. Zeng X., Winter D.B., Kasmer C. *et al.* (2001). DNA polymerase eta is an A-T mutator in somatic hypermutation of immunoglobulin variable genes. *Nat Immunol* **2**: 537–541.
114. Zhang H., Lawrence C.W. (2005). The error-free component of the RAD6/RAD18 DNA damage tolerance pathway of budding yeast employs sister-strand recombination. *Proc Natl Acad Sci USA* **102**: 15954–15959.
115. Zhang Y., Yuan F., Wu X. *et al.* (2001). Response of human DNA polymerase iota to DNA lesions. *Nucleic Acids Res* **29**: 928–935.

116. Zhang Y., Yuan F., Wu X. *et al.* (2000). Preferential incorporation of G opposite template T by the low-fidelity human DNA polymerase iota. *Mol Cell Biol* **20**: 7099–7108.
117. Ziv O., Geacintov N., Nakajima S. *et al.* (2009). DNA polymerase zeta cooperates with polymerases kappa and iota in translesion DNA synthesis across pyrimidine photodimers in cells from XPV patients. *Proc Natl Acad Sci USA* **106**: 11552–11557.

Chapter 7

Regulatory Mechanisms of AID Function

Almudena R. Ramiro[1] *and Javier M. Di Noia*[2]

[1]*Spanish National Cancer Research Centre (CNIO)*
Madrid, Spain
E-mail: arodriguezr@cnio.es

[2]*Institut de Recherches Cliniques de Montréal, 110 Av des Pins Ouest*
H2W 1R7, Montréal, Quebec, Canada
E-mail: javier.di.noia@ircm.qc.ca

A number of transcriptional, posttranscriptional and posttranslational mechanisms are known to regulate AID expression and function. Here we first overview the signaling pathways leading to AID expression in B cells *in vitro*, which mostly respond to surface receptors that include cytokine receptors such as IL4 and TGFβ, CD40 ligation or Toll-like receptors. Integration of these signals promotes the activation of *Aicda* (the gene encoding AID) transcription through various transcription factors, including the ubiquitous Sp1 or Sp3, lymphocyte-specific Pax5, E2A or HoxC4 as well as NFκB and hormone receptors. AID expression levels seem to be also subject to strict control, as evidenced by the functional impact of AID gene dose and AID overexpression in the various outcomes of its activity. At least two microRNAs (miR-155 and miR-181b) have been shown to contribute to this quantitative regulation. In addition, alternatively spliced variants of AID have been detected in malignant and healthy B cells, whose functional relevance remains unresolved. Finally, AID activity is exquisitely controlled by nuclear export, nuclear import, cytoplasmic retention and protein stability, which cooperatively restrict its nuclear concentration. The interconnections and relevance of this plethora of mechanisms remain a fascinating issue that is far from being understood.

7.1 Introduction

Activation-induced-deaminase (AID) plays a pivotal role in the secondary diversification of antibodies. AID initiates somatic hypermutation (SHM), class switch recombination (CSR) and immunoglobulin gene conversion (GCV) by deaminating cytosine residues in immunoglobulin genes. The essential role of AID in the humoral immune response is underscored by the immune deficiency entailed by mutations in *Aicda*, the gene encoding AID. However, AID-mediated deamination is also the initiating event for the occurrence of chromosomal translocations with lymphomagenic potential. It is expected that a number of regulatory mechanisms should be responsible for ensuring the efficient activity of AID in antibody diversification while preventing DNA mutations and genomic instability. In this chapter we discuss the transcriptional and posttranscriptional regulation of AID.

7.2 Transcriptional Regulation of AID Gene Expression

AID was originally believed to be exclusively expressed in activated B cells. However, recent reports have shown AID expression in other B cell subsets, and more strikingly, outside of the B cell lineage. In this section we will cover the pathways known to lead to AID expression, the transcription factors directly involved in promoting AID expression, including hormone-mediated regulation, and the haploinsufficient nature of AID.

7.2.1 *Expression of AID in and outside B cells*

AID is present in all vertebrates, spanning several hundred million years of evolution (Rogozin *et al.*, 2007), so considerable variation of its regulation can be expected among species. For instance, there are animals, such as birds, which use AID to diversify their primary repertoire (Reynaud *et al.*, 1987; Arakawa *et al.*, 2002). This requires the expression of AID in primary lymphoid tissues in an antigen-independent and developmentally-regulated fashion (Withers *et al.*,

2005), which most likely responds to stimuli different from those that induce AID during antigen-dependent antibody diversification in peripheral lymphoid tissues. In addition, it is now clear that AID can be induced outside of the B cell compartment. Triggers for this induction include pathological stimuli such as infections or inflammatory stimuli (Chiba and Marusawa, 2009; Gourzi *et al.*, 2007), but also physiological stimuli like estrogen, which is probably why AID is expressed in mouse ovaries (MacDuff *et al.*, 2009; Pauklin *et al.*, 2009). In this section, we will focus on the regulation of AID in human and mouse germinal center B cells, and briefly refer to the evidence of AID induction in some non-lymphoid tissues of these species.

The expression of AID within the murine and human B cell lineage is tightly regulated. Independent analyses of different B cell subpopulations and the use of reporter transgenic mice all agree in that, under physiological conditions, AID is mostly expressed during the germinal center reaction in secondary lymphoid organs (Cattoretti *et al.*, 2006; Crouch *et al.*, 2007; Greeve *et al.*, 2003; Muramatsu *et al.*, 1999; Pasqualucci *et al.*, 2004), specifically in centroblasts (Caron *et al.*, 2009; Cattoretti *et al.*, 2006; Greeve *et al.*, 2003), although it has also been observed in some extrafollicular proliferating B cells in human tonsils (Cattoretti *et al.*, 2006). Two reports have shown that AID expression can also be detected by sensitive RT-PCR in differentiating immature B cells in the bone marrow, probably at levels too low to be observable in reporter strains (Han *et al.*, 2007; Mao *et al.*, 2004). The functional relevance of AID expression at this location remains under debate, although it has been suggested that it could play a role in preimmune repertoire diversification and/or receptor editing (Han *et al.*, 2007; Mao *et al.*, 2004).

Several pathways are known that lead to AID induction *in vitro*. However, little is known about this *in vivo*, let alone the differences that may exist between the stimuli that induce AID in naïve B cells after activation versus centroblasts undergoing cycles of mutation and selection during the germinal center reaction or memory B cells that might be recruited back to germinal centers.

7.2.2 *Signal transduction pathways leading to Aicda induction*

Consistent with the different contexts in which AID can be induced, several pathways can lead to AID expression. AID was originally isolated from the murine CH12F3 cell line stimulated to undergo class switching using IL-4, CD40 agonist and TGFβ1 (Muramatsu *et al.*, 1999), which already suggested that the signaling pathways downstream from the corresponding receptors would induce *Aicda* transcription. In fact, signaling through the IL-4R and CD40 is synergistic for inducing AID in mouse and human B cells (Dedeoglu *et al.*, 2004; Muramatsu *et al.*, 1999; Zhou *et al.*, 2003). These pathways are likely to partially mimic B cell/T cell interactions. The actual *in vivo* context is likely to be more complex than this given the multiple cellular interactions, including those from antigen-presenting cells (Wu *et al.*, 2008b), occurring during B cell-activation and the germinal center reaction. In addition, AID induction is transient after B cell-activation (Greeve *et al.*, 2003; Muramatsu *et al.*, 1999) so there must be stimuli to downregulate the gene after a short while. Indeed, a mechanism to stop diversification, once a certain affinity for the antigen has been achieved, makes teleological sense. Although the evidence is still scarce, AID downregulation seems to require signalling through the BCR. This is supported by the effect of anti-IgM added to stimulated murine B cells, which delays AID induction (Jabara *et al.*, 2008), and BCR cross-linking, which downregulates AID through a Ca++/calmodulin-dependent pathway (Hauser *et al.*, 2008).

AID is also expressed during T cell-independent B cell responses (Chen *et al.*, 2009; Crouch *et al.*, 2007). The engagement of TACI by BAFF and/or APRIL from dendritic cells, probably combined with TGFβ1 signalling, seems to be the inducing stimulus for AID (Chen *et al.*, 2009; Litinskiy *et al.*, 2002). TLR agonists such as LPS and CpG can induce AID in murine splenic and bone marrow B cells (Gourzi *et al.*, 2007; Muramatsu *et al.*, 1999), suggesting another T cell-independent induction pathway of yet unclear physiological significance. It is possible that this relates to the role of AID in diversifying the primary repertoire of some species in the gut-associated lymphoid tissues (Lanning *et al.*, 2000; Reynaud *et al.*, 1987; Reynaud *et al.*, 1989;

Reynaud *et al.*, 1995; Withers *et al.*, 2005; Yang *et al.*, 2005), or the ancient role of AID in creating diversity in species without VDJ recombination (Rogozin *et al.*, 2007). It also remains possible that this alternative pathway of AID induction is related to a proposed original role in restricting foreign DNA (MacDuff *et al.*, 2009; Petersen-Mahrt *et al.*, 2009), which the ability of some viral infections to induce AID may support. This effect has been documented for instance for HCV (Machida *et al.*, 2004) and for Abelson MLV (Gourzi *et al.*, 2006), but the signalling pathway is unknown except for the fact that the TLR signalling is not necessary for AID induction by Abelson MLV (Gourzi *et al.*, 2007). In fact, there seems to be a common theme linking AID expression to proinflammatory stimuli such as TGFβ1 (Muramatsu *et al.*, 1999) or TNFα (Pauklin *et al.*, 2009). The many examples of AID induction in cancer-associated inflammatory states (Chiba and Marusawa, 2009) further supports this link and, together with the viral induction of AID and the role of the TLR, suggests an important role for NFκB in inducing AID.

7.2.3 *Transcription factors inducing AID*

The *Aicda* gene is well conserved between mice and humans. The promoter, transcription start site and a 3′ enhancer have been mapped (Crouch *et al.*, 2007; Yadav *et al.*, 2006). The *Aicda* promoter and intron 1 regulatory sequences are quite similar between mice and humans (Crouch *et al.*, 2007; Gonda *et al.*, 2003; Park *et al.*, 2009) and include binding sites for a number of transcription factors (Dedeoglu *et al.*, 2004; Gonda *et al.*, 2003; Park *et al.*, 2009; Pauklin and Petersen-Mahrt, 2009; Pauklin *et al.*, 2009; Sayegh *et al.*, 2003; Yadav *et al.*, 2006). Whether the most proximal promoter can provide B cell specificity is unclear (Gonda *et al.*, 2003; Yadav *et al.*, 2006) but most likely this would require additional *cis*-acting elements (Crouch *et al.*, 2007). It is not surprising that numerous transcription factors, both ubiquitous as well as more B cell- restricted ones bind *Aicda* promoter and intron 1, suggesting a complex transcriptional regulation.

As suggested by the type of stimuli inducing AID, NFκB plays a prominent role in this induction. NFκB binds to the *Aicda* promoter in response to CD40 signalling and is crucial for inducing its expression as $p50^{-/-}$ B cells lose the synergistic effect between IL-4 and CD40L (Dedeoglu *et al.*, 2004). This may explain why $p50^{-/-}$ mice are deficient for isotype switching (Snapper *et al.*, 1996). NFκB is also downstream of TLR signalling, which seems to be one way in which viral or bacterial infections as well as inflammation can induce AID (Chiba and Marusawa, 2009; Gourzi *et al.*, 2007; Matsumoto *et al.*, 2007). In addition, the TLR-independent induction of AID by Abelson MLV infection requires NFκB (Gourzi *et al.*, 2007). It is likely that the T cell-independent signaling through the TACI agonists also feeds into the NFκB pathway to activate *Aicda*. A second important factor for inducing AID is STAT6, which binds to *Aicda* promoter in response to IL-4 (Dedeoglu *et al.*, 2004). Furthermore, *Aicda* transcription is inhibited by a STAT6 dominant negative form in reporter cells, indicating the functional relevance of this promoter binding (Zhou *et al.*, 2003). The nuclear translocation of dimeric STAT6 results from its phosphorylation by Janus kinases, which are associated with the IL-4R (Reich and Liu, 2006). This step can be inhibited by the tyrosine phosphatase CD45 in B cells, which thus acts as a repressor of AID (Zhou *et al.*, 2003).

Among the more cell-type specific transcription factors, the B cell lineage-specifier Pax5 (Cobaleda *et al.*, 2007) has been implicated in *Aicda* transcription. Pax5 binds to and activates *Aicda* as shown by chromatin immunoprecipitation and *in vitro* assays (Gonda *et al.*, 2003; Park *et al.*, 2009). There is certainly a positive correlation between expression of Pax5 and AID (Gonda *et al.*, 2003; Nera *et al.*, 2006). It is therefore possible that the repression of Pax5 by Blimp-1 (Lin *et al.*, 2002) contributes to shutting off *Aicda* transcription during B cell differentiation to the plasmablast stage. However, there is no agreement on whether Pax5 would bind *Aicda* directly (Gonda *et al.*, 2003) or as a complex with the ubiquitous factors Sp1 or Sp3 (Park *et al.*, 2009; Yadav *et al.*, 2006). The E2A transcription factors, E12 and E47, which are also important for B cell lineage commitment (Cobaleda *et al.*, 2007; Sugai *et al.*, 2004), have also been reported to induce AID (Sayegh *et al.*, 2003), but this finding could not be confirmed by others (Gonda *et al.*, 2003). Nevertheless, both groups agreed in that Id2 and Id3, which are known antagonists of E2A (Sugai *et al.*, 2004), negatively regulate AID expression, thus suggesting a role for E2A proteins in inducing AID. Supporting this possibility, inhibition of E2A seems to be responsible for AID downregulation after BCR signalling (Hauser *et al.*, 2008). It is still possible that Id2 actually antagonizes Pax5 (Gonda *et al.*, 2003; Sugai *et al.*, 2004), but this might be an indirect effect. The latest addition to the list of transcription factors positively regulating *Aicda* has been HoxC4, which acts synergistically with Oct1/Oct2 and binds to the *Aicda* promoter (Park *et al.*, 2009). HoxC4 expression seems to be restricted to germinal centers in adult mice and humans and to be induced by B cell-activating stimuli, presumably as a consequence of CD40 signalling (Park *et al.*, 2009). This study also detected recruitment of Oca-B to *Aicda* but its effect was not characterized.

One further layer of complexity has been recently added to the type of transcription factors regulating Aicda gene expression: a likely role of the sex hormone receptors. Estrogen and progesterone both affect AID expression with opposite effects. Exposure of B cells to estrogen results in the binding of the estrogen receptor (ER) to *Aicda*, transcriptional activation and AID activity (Pauklin *et al.*, 2009). The ER pathway is proposed to interact in some way with the NFκB pathway since costimulation with TNFα and estrogen has synergistic effects on AID expression (Pauklin *et al.*, 2009; Petersen-Mahrt *et al.*, 2009). On the other side, the progesterone receptor binds to and represses *Aicda* transcription (Pauklin and Petersen-Mahrt, 2009). Estrogen can also upregulate AID in non-lymphoid responsive tissues and does so in ovaries and breast (Pauklin *et al.*, 2009). Whether this serves some kind of physiological purpose or is just a relic of a previous function of AID remains speculative, but it certainly raises the possibility that it may contribute to oncogenesis in those tissues (Petersen-Mahrt *et al.*, 2009). More importantly perhaps, it is well known that sex hormones do have an effect on adaptive immunity (Beagley and Gockel, 2003) and humoral immunity in particular (Beagley and Gockel, 2003; Kincade *et al.*, 2000). In fact, the estrogen receptors are expressed and functional during B cell development including the germinal center stages (Grimaldi *et al.*, 2002; Grimaldi *et al.*, 2005). It is therefore likely that sex hormones do participate in modulating the expression of AID *in vivo*, a nice example of the well established, albeit less well understood at the molecular level, connection between the endocrine and immune systems.

The picture emerging from the role of the various signaling pathways and transcription factors impinging on AID expression is complex (Fig. 7.1). Although some relationships between pairs of these factors have been suggested, considerable work will be necessary to have a clear

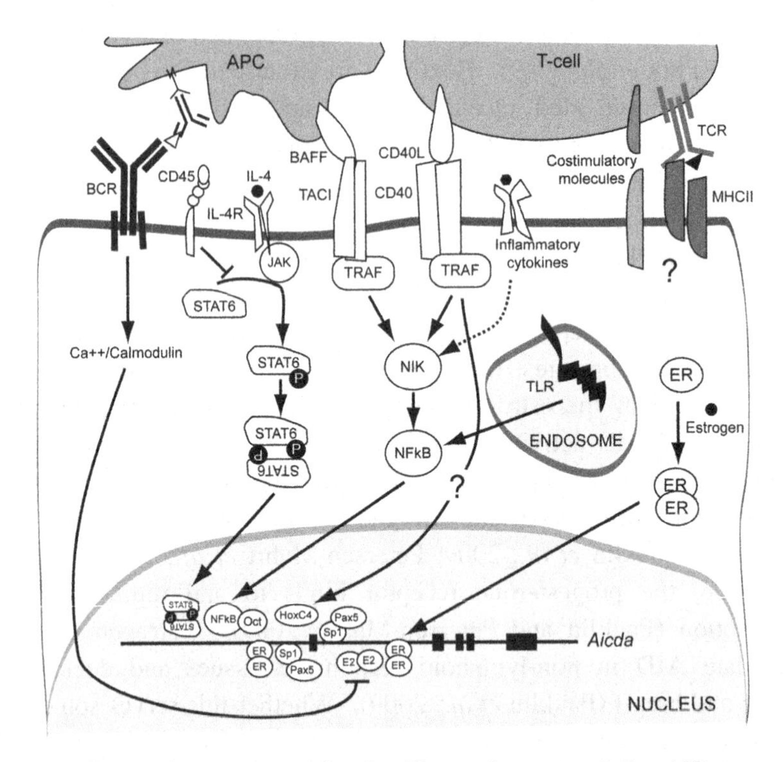

Figure 7.1. **Schematic summary of the signalling pathways regulating AID expression in B cells.** Receptors on the B cell surface that upon engagement of their ligand are known to affect AID induction are depicted. Although most of the evidence for their involvement relies on *in vitro* stimulation experiments of purified B cells and B cell lymphoma lines, the potential role of antigen-presenting cells (APC) and T lymphocytes in contributing some of these stimuli is shown. A simplified scheme of the signalling pathways downstream from each receptor in the cytoplasm is shown and the transcription factors that they activate or repress are drawn on top of the *Aicda* gene in the nucleus. Arrows indicate positive stimulation and T-lines indicate negative regulation. Unclear pathways or when a role for the molecules drawn is expected but has not been experimentally addressed are indicated by question marks.

understanding of how all of them are integrated to achieve the right amount of *Aicda* transcription in response to physiological stimuli *in vivo*.

7.2.4 *AID haploinsufficiency*

Once AID expression has been induced by either one of the above mentioned pathways, the question arises of whether the control of AID physiological levels could be a determinant for its activity. Several recent reports have provided evidence that this is indeed the case. $AID^{+/-}$ B cells express roughly half the amount of AID both at the mRNA and protein levels (McBride *et al.*, 2008; Sernandez *et al.*, 2008; Takizawa *et al.*, 2008). This decrease in AID expression results in a diminished ability of $AID^{+/-}$ B cells to undergo CSR after *in vitro* stimulation with different combinations of cytokines. The extent of CSR impairment of $AID^{+/-}$ differs among different reports; after three days of stimulation in the presence of LPS and IL4 the rate of CSR of $AID^{+/-}$ compared to $AID^{+/+}$ cells ranged from 90 to 40%, with the most pronounced reduction observed in two studies using respective mice in the Balb/c background, suggesting a strain-specific contribution to the observed phenotype. The reduction of CSR observed *in vivo* in immunized Balb/c mice was milder than after *in vitro* stimulation, probably as a result of the selective events taking place in germinal centers (Takizawa *et al.*, 2008). A reduction in the mutation load has also been observed in mice harbouring a single functional AID allele. This reduction affects both the frequency of mutation in the Sμ region after *in vitro* LPS+IL4 stimulation and the J_H4 intronic region of Peyers patch B cells and of germinal center B cells from immunized mice.

AID haploinsufficiency is also obvious in the rate of chromosome translocations generated in different *in vivo* and *in vitro* systems; pristane treated Bcl-xL transgenic $AID^{+/-}$ mice (Takizawa *et al.*, 2008) and IL6tg $AID^{+/-}$ (Sernandez *et al.*, 2008) accumulate fewer c-myc/IgH translocations than $AID^{+/+}$ littermates, and LPS+IL4 stimulated B cells from $AID^{+/-}$ mice contain fewer DNA double-strand breaks in switch regions (Takizawa *et al.*, 2008) and fewer c-myc/IgH translocations (Sernandez *et al.*, 2008) than control animals. Importantly, although

Bcl-xL transgenic AID$^{+/-}$ mice developed tumors after pristane treatment, their incidence was significantly reduced when compared to their AID wild-type counterparts (Takizawa *et al.*, 2008). In addition, AID$^{+/-}$ mice exhibit a delayed production of high affinity antibodies against pathogens and in the development of autoimmune disease in a lupus prone model (Jiang *et al.*, 2007; Jiang *et al.*, 2009). These latter findings highlight the relevance of AID gene dose for B cell neoplasia generation and autoimmune susceptibility *in vivo.* Together these results indicate that physiological AID expression levels are limiting for its function, probably allowing an efficient diversification of antibodies while preventing unwanted DNA lesions. However, the interplay between this and other AID regulatory mechanisms can be unexpectedly complex.

AID subcellular localization is exquisitely controlled (see below), and the fraction of nuclear protein is strictly limited. The fact that a reduction of total AID protein has an almost linear effect on its activity suggests that the fraction of nuclear AID is constant. In addition, AID function is regulated by phosphorylation by at least two independent residues, S38 and T140 (see Chapter 4 of this monograph). Knock-in mice harbouring alanine mutations in these residues display an impairment in CSR and SHM, however this effect increases dramatically in haploinsufficient mice (mice whose only *Aicda* copy encodes the unphosphorylatable mutant; Cheng *et al.*, 2009; McBride *et al.*, 2008); this synergistic effect suggests that when AID levels drop below a functional threshold, phosphorylation at these residues becomes more critical for its activity. Finally, it is worth noting that the effects of AID overexpression do not faithfully mirror those of AID deficit. Surpassing AID physiological levels both *in vitro* and *in vivo* results in an increase of CSR and in the rate of chromosomomal translocations (Dorsett *et al.*, 2008; Ramiro *et al.*, 2006; Robbiani *et al.*, 2009; Sernandez *et al.*, 2008; Teng *et al.*, 2008). However, mice expressing over-physiological levels of AID do not seem to develop B cell lymphomas (Dorsett *et al.*, 2008; Muto *et al.*, 2006). A plausible explanation for this B cell resistance to transformation has been proposed very recently (Robbiani *et al.*, 2009). In this report, transgenic mice were generated that allow B cell-specific overexpression of AID, which results in enhanced SHM, CSR and c-myc/IgH translocations (Robbiani *et al.*, 2009). Interestingly, while

AIDtg mice do not develop lymphomas up to 40 weeks of age, transgenic AID expression in a p53-deficient background leads to the development of mature B cell tumors, with a median survival of 24 weeks (Robbiani *et al.*, 2009). These results indicate that AID overexpression by itself is not sufficient in promoting B cell lymphomagenesis, but instead tumor-suppressor pathways prevent AID-mediated B cell transformation. In contrast to a previous report (Ramiro *et al.*, 2006), p53 does not seem to preclude the generation of c-myc/IgH fusions in this model, and is more likely to be activated by oncogenic stress (Robbiani *et al.*, 2009). It is possible that again AID expression levels may be important in determining the contribution of other regulatory pathways. In this particular case, one can speculate that the lesions promoted by physiological levels of AID trigger a p53-dependent checkpoint that prevents the generation of c-myc/IgH translocations, while deregulated AID gives rise to an overwhelming amount of damage that relies on additional tumor-suppressor functions of p53. Overall, these observations suggest on the one hand an intertwined network of mechanisms regulating AID function, and on the other, the existence of a post-AID regulatory layer that protects B cells from transformation in the event that accumulation of AID-promoted DNA lesions cannot be prevented.

7.3 Posttranscriptional Regulation of mRNA Levels

7.3.1 *Regulation of AID expression by microRNAs*

In recent years microRNAs have emerged as a novel layer of control of gene expression that plays a crucial role in the regulation of numerous biological processes (Bushati and Cohen, 2007). MicroRNAs are small molecules (21 to 23 nucleotide long) of non-coding single-stranded RNA that can regulate gene expression in a posttranscriptional (by decreasing mRNA stability) or posttranslational (by impairing protein translation) fashion. This control is driven by partial sequence complementarity with their target mRNAs. The major determinant for microRNA specificity for its mRNA targets is driven by a perfect (or quasi-perfect) match in the so-called "seed" region, a 6–8 nucleotide long sequence at the 5′ end of the microRNA with a complementary sequence that is very often located in the 3′ UTR region of the mRNA target (He and Hannon, 2004). A number of microRNAs have been shown to be involved in the regulation of different aspects of the immune system, including miR-17-92, miR-150, miR-181 and miR-155 (Xiao and Rajewsky, 2009).

In the search for microRNAs able to regulate AID expression, two laboratories independently found a potential binding site for miR-155 in the 3′ UTR of AID mRNA. miR-155 is expressed in hematopoietic cells and upregulated in B and T cells upon activation. Previous *in vivo* gain- and loss-of-function analysis had shown that miR-155 plays a positive regulatory role for the germinal center reaction and the generation of plasma cells (Rodriguez *et al.*, 2007; Thai *et al.*, 2007; Vigorito *et al.*, 2007). To test whether miR-155 could be a regulator of AID expression, the Nussenzweig and Papavasiliou laboratories undertook a novel *in vivo* approach based on introducing mutations in the AID 3′ UTR to abolish the potential binding of miR-155 and generating knock-in (AID155-KI) or BAC-transgenic mouse strains (AID155-tg), respectively (Dorsett *et al.*, 2008; Teng *et al.*, 2008). These models therefore allow insulating AID regulation by miR-155 from other miR-155-mediated functions. Importantly, in both cases ablation of miR-155 binding site resulted in an increase of AID protein levels, directly proving that AID mRNA is indeed a target for miR-155. This upregulation was restricted to activated B cells as AID expression was not observed in other cell lineages or B cell developmental stages (Dorsett *et al.*, 2008; Teng *et al.*, 2008). Elevated AID protein levels in AID155-KI and AID155-tg resulted in a substantial increase in the frequency of class-switched cells generated *in vitro*. In contrast, the mutation frequency was not significantly altered in B cells from immunized animals (Dorsett *et al.*, 2008; Teng *et al.*, 2008) or in switch regions after B cell stimulation with LPS and IL4 (Dorsett *et al.*, 2008). In addition, one of the studies reported a three- to six-fold increase in the frequency of c-myc/IgH chromosome translocations in B cells upon activation while miR155$^{-/-}$ B cells showed an even larger propensity to these events (15-fold increased compared to controls; Dorsett *et al.*, 2008). This intriguing differential regulation of AID overexpresion for CSR/c-myc-IgH translocations versus SHM was however not observed in a transgenic AID overexpression mouse model recently reported by Robbiani and colleagues, where all three outputs of AID activity were similarly enhanced (Robbiani *et al.*, 2009). Together these findings have unveiled a new, microRNA-mediated mechanism regulating AID expression. However, the regulation of the germinal center reaction in general by miR-155 seems to involve additional layers of complexity. Although increased AID expression had also been observed in miR-155$^{-/-}$ B cells (Dorsett *et al.*, 2008; Vigorito *et al.*, 2007), these animals show an impaired generation of germinal center cells (Rodriguez *et al.*, 2007; Thai *et al.*, 2007; Vigorito *et al.*, 2007), which suggests that in miR155$^{-/-}$ mice the effect of AID upregulation in CSR is counteracted by other miR-155 regulated genes, including Pu.1 (Vigorito *et al.*, 2007). These seemingly contradictory effects of miR-155 might achieve a suitable balance of positive and negative regulatory mechanisms for germinal center formation. An alternative approach addressed the identification of microRNAs involved in the regulation of CSR by a functional screen of a microRNA library in primary B cells. This screen allowed the identification of miR-181b as a microRNA that can reduce the efficiency of CSR to about half of normal levels (de Yebenes *et al.*, 2008). A combination of expression profiling, bioinformatics search for seed sequences and reporter assays showed that miR-181b promotes AID downregulation by directly targeting its 3′ UTR. miR-181b expression is regulated in the context of B cell activation, but in contrast to miR-155, miR-181b levels drop upon stimulation of B cells (de Yebenes *et al.*, 2008). Together, these data reinforce the idea that AID expression is subject to stringent regulatory mechanisms and suggest that independent microRNA-based mechanisms might have evolved to negatively regulate AID expression at different developmental stages.

7.3.2 *AID alternative splicing*

Generation of alternatively spliced mRNA forms is a common mechanism responsible for the regulation of protein activity. The *Aicda* gene comprises five exons. Exon 1 (aa 1–3) and exon 2 (aa 4–52) have been reported to contain most of the nuclear localization signal (Ito *et al.*,

2004; Patenaude *et al.*, 2009) and to be required for SHM (Shinkura *et al.*, 2004). Exon 3 spans residues 53–142 and contains the cytosine deaminase domain (aa 54–94) and part of an Apobec-like domain that also spans the whole of exon 4 (aa 143–181). Finally, exon5 (aa 182–198) contains a well-defined nuclear export signal (see below) and is essential for CSR (Shinkura *et al.*, 2004; Barreto *et al.*, 2003; Ta *et al.*, 2003).

In 2003, two alternatively spliced mRNA isoforms were identified in patients with B cell chronic lymphocytic leukemia (B-CLL). These isoforms differed from the canonical AID mRNA (exon 1–5) in either the skipping of exon 4 (AIDΔexon4) or the inclusion of intron 4. Both isoforms were predicted to give rise to truncated proteins (147 and 187 residues long, respectively; McCarthy *et al.*, 2003). These forms were also detected in mantle cell lymphoma cells (Babbage *et al.*, 2004) and an additional one containing a small insertion between exons 3 and 4 was identified in B-CLL cells (Albesiano *et al.*, 2003). More recently, Jelinek and colleagues showed that spliced variants were not specific to transformed cells, as they could detect AIDΔexon4 in tonsillar and memory B cells but not in naive B cells from human peripheral blood (Wu *et al.*, 2008a). These alternative isoforms are expressed at very low levels and apparently contribute to the total AID mRNA in a much lower extent than they do in malignant cells. However, a single cell PCR assay with isoform specific primers would be required to address this unequivocally. Likewise, low levels of AIDΔexon3Δexon4 are present in murine B cells, indicating that at least some of this alternative processing is conserved between mouse and man (Wu *et al.*, 2008a). The authors addressed the functionality of AIDΔexon4a, AIDΔexon4 and AIDΔexon3Δexon4 and found that none of them could reconstitute CSR after overexpression in murine B cells with an AID$^{-/-}$ background. In contrast, they reported that AIDΔexon4a and AIDΔexon4 display mutation activity when expressed in 70Z/3 cells and analysed by a GFP reversion assay (Wu *et al.*, 2008a). However, these latter results have been challenged by the finding that none of the three splice variants were able to deaminate an oligonucleotide *in vitro* (van Maldegem *et al.*, 2009). A recent report has managed to detect protein expression from AID mRNA variants in BCR-ABL1+ ALL patients, further strengthening the case for an as yet unknown function for these alternatively spliced forms (Iacobucci *et al.*, 2010).

In summary, the expression of alternatively spliced forms of AID is an interesting issue that deserves further attention. It is tempting to speculate that inactive truncated forms of AID potentially retaining dimerization ability could serve as dominant negative regulators of CSR, SHM or both.

7.4 Posttranslational Control of AID

The next level for regulating the amount of available AID once it is translated seems particularly focused on restricting the amount of the enzyme in the nucleus. Both the regulation of AIDs sub-cellular localization and AID protein stability are interrelated mechanisms that control its activity posttranscriptionally. In addition, modulation of AID enzymatic activity and its recruitment to single-stranded DNA by Ser/Thr phosphorylation, as well as binding to the transcription machinery and mRNA-processing cofactors have been described and are reviewed in chapter 4.

7.4.1 *AID subcellular localization and stability*

One of the most conspicuous and earliest recognized instances of posttranslational regulation of AID is its subcellular compartmentalization. It was noted early on that an AID-GFP fusion protein appeared to be exclusively cytoplasmic by fluorescence microscopy, despite being able to increase the Ig gene mutation rate in Ramos B cells (Rada *et al.*, 2002). The demonstration that AID is in fact a nucleus-cytoplasm shuttling protein (Brar *et al.*, 2004; Ito *et al.*, 2004; McBride *et al.*, 2004) reconciled this observation with the function of AID as a DNA deaminase in the nucleus. The current evidence suggests that there are three mechanisms determining AID compartmentalization: nuclear export, nuclear import and cytoplasmic retention. The balance of these three mechanisms is shifted to favor the cytoplasmic localization that is observed in steady state.

Nuclear export was the first of these mechanisms to be characterized. The last 10–11 residues of AID bear a self-sufficient canonical leucine-rich nuclear export signal (NES) recognized by the exportin CRM1, which mediates the exit of AID from the nucleus. This was first demonstrated by the nuclear accumulation of AID upon treatment with the CRM1-specific inhibitor leptomycin-B (Brar *et al.*, 2004; Ito *et al.*, 2004; McBride *et al.*, 2004). So far, this has only been shown for tagged and overexpressed versions of AID. Based on the fact that C-terminal truncations of AID perform SHM but are unable to catalyze class switch recombination (CSR; Barreto *et al.*, 2003; Shinkura *et al.*, 2004; Ta *et al.*, 2003), it was proposed that nuclear export might be required for CSR. However, recent data showing that a heterologous NES cannot substitute for the AID NES to promote CSR suggest that this region rather participates in binding some CSR-specific cofactor (Geisberger *et al.*, 2009).

The nuclear import pathway of AID has been a matter of debate, with both passive diffusion and the existence of a bipartite nuclear localization signal (NLS) being proposed (Brar *et al.*, 2004; Ito *et al.*, 2004; McBride *et al.*, 2004). Experimental evidence that AID is actively imported into the nucleus was recently reported (Patenaude *et al.*, 2009). However, rather than a linear self-sufficient NLS, a conformational NLS was proposed. This NLS is mostly constituted by basic residues within the first 50 amino acids of AID but it may also contain residues located further downstream and/or require the folding of AID to acquire the right spatial configuration that allows recognition of the key NLS residues by importin-α adaptors (Patenaude *et al.*, 2009). Although, the available evidence suggests that AID nuclear import proceeds through the canonical importin-α/importin-β (Lange *et al.*, 2007), this remains to be formally demonstrated.

The work demonstrating that AID was actively imported into the nucleus unexpectedly revealed that the diffusion of AID from the cytoplasm was impaired, which suggested the existence of a mechanism for the cytoplasmic retention of AID. This was inferred by the failure of AID (both tagged and untagged) to passively diffuse into the nucleus in conditions in which nuclear import and export were inhibited by different means (Patenaude *et al.*, 2009). In the absence of any transport mechanism, AID should equilibrate across the nuclear pores by passive diffusion according to the mass action law. Since it remained cytoplasmic it must be concluded that some retention mechanism exists. The domain of AID found to mediate cytoplasmic retention overlaps with the NES, which complicates its functional characterization. However, mutations separating the function of each of these mechanisms were described, thus confirming that they are distinct (Patenaude *et al.*, 2009). The molecular basis of AID cytoplasmic retention is unknown. It is also unclear how much nuclear export and cytoplasmic retention contribute to the nuclear exclusion of AID in steady state. Most of the work done on AID compartmentalization has used overexpressed protein because of the lack of appropriate reagents to study endogenous AID. The limited evidence available on endogenous AID localization shows that inhibition of nuclear export by leptomycin B does not suffice to cause significant nuclear localization of AID in Ramos B cells (Patenaude *et al.*, 2009). It is therefore possible that the contribution of cytoplasmic retention to AID nuclear exclusion is important, but this requires further work to be substantiated.

Regardless of the mechanism by which the compartmentalization of AID is achieved, it is clear that the nuclear exclusion of AID restrains its physiological activity in antibody diversification and probably its ability for off-target mutations. Indeed, AID variants with their C-terminal region truncated or mutated, which are constitutively nuclear, show exacerbated SHM, Ig gene conversion and mutation of non-Ig genes (Aoufouchi *et al.*, 2008; Barreto *et al.*, 2003; McBride *et al.*, 2004). Interestingly, a recent report suggests that, although these variants are unable to perform CSR (Shinkura *et al.*, 2004; Barreto *et al.*, 2003), they have increased ability to cause IgH/cMyc translocations (Doi *et al.*, 2009). Since mutations in the NLS also abrogate CSR (Patenaude *et al.*, 2009) it is possible that active import is important in overcoming the combined action of export and retention to allow AID function. The common caveat to all these experiments is the possibility that the mutants employed also affect the binding of as of yet uncharacterized factors.

The presence of AID in the nucleus seems to be restricted also by protein destabilization. There is clear evidence showing that the half-life of nuclear AID is significantly shorter than that of cytoplasmic AID (Aoufouchi *et al.*, 2008). After nuclear import, AID is polyubiquitylated and degraded by the proteasome in the nucleus (Aoufouchi *et al.*, 2008). The exact pathway of degradation is unknown and no specific lysine residue seems to be the target for the polyubiquitylation. Recently, an increase of nuclear AID-YFP in DT40 cells during the G1 stage of the cell-cycle was reported (compared to the amounts observed in S phase), which was apparently due to preferential destabilization of AID in the S phase (Ordinario *et al.*, 2009), suggesting that nuclear destabilization might be controlled temporally. Additional work is necessary to reveal the molecular details of the pathways determining AID stability.

The C-terminus of AID has been proposed to modulate AID stability independently of its role in nuclear export. Indeed, mutations that did not affect the NES but caused reduced CSR efficiency were found to be destabilizing, as was the case for AID-carrying replacements of its own NES by heterologous NES, which were excluded from the nucleus but were still CSR-deficient (Geisberger *et al.*, 2009). However, the finding that the NES and a motif important for cytoplasmic retention overlap (Patenaude *et al.*, 2009) could reconcile these observations with the work of Aoufouchi and colleagues (Aoufouchi *et al.*, 2008): The increased turn-over of the mutant proteins lacking CSR activity may be the consequence of reduced cytoplasmic retention and therefore increased nuclear import.

7.5 Integration of AID Regulation: The Outstanding Questions

There is a striking complexity in the mechanisms that regulate the abundance of AID, all of which seem to be necessary since ablation or inhibition of any of them leads to more diversification and off-target activity, as we have reviewed here. The interactions between all these pathways are essentially unknown. However, several observations suggest that there is substantial interplay not only between the mechanisms regulating *Aicda* gene transcription or those regulating the encoded protein, but also between the different strata of AID regulation.

For instance, it is well known that stimulation of murine B cells *in vitro* with various cytokines induces AID and robust CSR, but not SHM. Is it possible that the signaling that induces antibody diversification can influence the targeting of AID to different regions of the Ig locus? Perhaps some factor(s) necessary for SHM are induced or activated only under certain conditions that are not mimicked by the stimuli used for CSR *in vitro*.

The interplay between AID subcellular localization and protein stability is very clear. However, at least three mechanisms ensure nuclear exclusion of AID: nuclear export, cytoplasmic retention and nuclear destabilization. What is the relative importance of each of these pathways? Do they all compete for the same pool of AID simply to achieve the desired level of AID protein? An alternative possibility is that they are important at different stages of the reaction (e.g. before or after arriving to the Ig loci). There is still some work required to resolve the issue of whether the bulk of cellular AID needs to translocate into the nucleus for antibody diversification (for instance at a particular stage of the cell-cycle) or whether the amount of nuclear AID is constant at all times. So far, the only evidence supporting the first possibility is the observation of a very small proportion of human tonsillar B cells with nuclear staining in immunohistochemistry studies (Cattoretti *et al.*, 2006). A constant ratio between nuclear and cytoplasmic AID is instead supported by the phenotype of haploinsufficient mice and by complementation experiments showing that AID variants that are very inefficient in entering the nucleus can still catalyze substantial diversification (Patenaude *et al.*, 2009; Shinkura *et al.*, 2004). However, how this equilibrium is achieved exactly is unknown. Finally, does phosphorylation or any other posttranslational modification (apart from polyubiquitylation) participate in regulating the subcellular localization or stability of AID? Much remains to be done about this.

7.6 Acknowledgements

Work in the laboratory of J.M.D.N. is funded by the Canadian Institutes of Health Research and The Cancer Research Society. J.M.D.N. is supported by a Canada Research Chair Tier 2. A.R.R. is funded by the Ramón y Cajal Program and by grants from the Ministerio de Ciencia e Innovación (SAF2007-63130), the Comunidad Autónoma de Madrid (DIFHEMAT-CM) and the European Research Council Starting Grant program (BCLYM-207844).

7.7 References

1. Albesiano E., Messmer B.T., Damle R.N. *et al.* (2003). Activation-induced cytidine deaminase in chronic lymphocytic leukemia B cells: expression as multiple forms in a dynamic, variably sized fraction of the clone. *Blood* **102**: 3333–3339.
2. Aoufouchi S., Faili A., Zober C. *et al.* (2008). Proteasomal degradation restricts the nuclear lifespan of AID. *J Exp Med* **205**: 1357–1368.
3. Arakawa H., Hauschild J., Buerstedde J.M. (2002). Requirement of the activation-induced deaminase (AID) gene for immunoglobulin gene conversion. *Science* **295**: 1301–1306.
4. Babbage G., Garand R., Robillard N. *et al.* (2004). Mantle cell lymphoma with t(11;14) and unmutated or mutated VH genes expresses AID and undergoes isotype switch events. *Blood* **103**: 2795–2798.
5. Barreto V., Reina-San-Martin B., Ramiro A.R. *et al.* (2003). C-terminal deletion of AID uncouples class switch recombination from somatic hypermutation and gene conversion. *Mol Cell* **12**: 501–508.
6. Beagley K.W., Gockel C.M. (2003). Regulation of innate and adaptive immunity by the female sex hormones oestradiol and progesterone. *FEMS Immunol Med Microbiol* **38**: 13–22.
7. Brar S.S., Watson M., Diaz M. (2004). Activation-induced cytosine deaminase (AID) is actively exported out of the nucleus but retained by the induction of DNA breaks. *J Biol Chem* **279**: 26395–26401.
8. Bushati N., Cohen S.M. (2007). microRNA functions. *Annu Rev Cell Dev Biol* **23**: 175–205.
9. Caron G., Le Gallou S., Lamy T. *et al.* (2009). CXCR4 expression functionally discriminates centroblasts versus centrocytes within human germinal center B cells. *J Immunol* **182**: 7595–7602.
10. Cattoretti G., Buttner M., Shaknovich R. *et al.* (2006). Nuclear and cytoplasmic AID in extrafollicular and germinal center B cells. *Blood* **107**: 3967–3975.
11. Chen K., Xu W., Wilson M. *et al.* (2009). Immunoglobulin D enhances immune surveillance by activating antimicrobial, proinflammatory and B cell-stimulating programs in basophils. *Nat Immunol* **10**: 889–898.
12. Cheng H.L., Vuong B.Q., Basu U. *et al.* (2009). Integrity of the AID serine-38 phosphorylation site is critical for class switch recombination and somatic hypermutation in mice. *Proc Natl Acad Sci USA* **106**: 2717–2722.

13. Chiba T., Marusawa H. (2009). A novel mechanism for inflammation-associated carcinogenesis; an important role of activation-induced cytidine deaminase (AID) in mutation induction. *J Mol Med* **87**: 1023–1027.
14. Cobaleda C., Schebesta A., Delogu A. *et al.* (2007). Pax5: the guardian of B cell identity and function. *Nat Immunol* **8**: 463–470.
15. Crouch E.E., Li Z., Takizawa M. *et al.* (2007). Regulation of AID expression in the immune response. *J Exp Med* **204**: 1145–1156.
16. de Yebenes V.G., Belver L., Pisano D.G. *et al.* (2008). miR-181b negatively regulates activation-induced cytidine deaminase in B cells. *J Exp Med* **205**: 2199–2206.
17. Dedeoglu F., Horwitz B., Chaudhuri J. *et al.* (2004). Induction of activation-induced cytidine deaminase gene expression by IL-4 and CD40 ligation is dependent on STAT6 and NFkappaB. *Int Immunol* **16**: 395–404.
18. Doi T., Kato L., Ito S. *et al.* (2009). The C-terminal region of activation-induced cytidine deaminase is responsible for a recombination function other than DNA cleavage in class switch recombination. *Proc Natl Acad Sci USA* **106**: 2758–2763.
19. Dorsett Y., McBride K.M., Jankovic M. *et al.* (2008). MicroRNA-155 suppresses activation-induced cytidine deaminase-mediated Myc-Igh translocation. *Immunity* **28**: 630–638.
20. Geisberger R., Rada C., Neuberger M.S. (2009). The stability of AID and its function in class-switching are critically sensitive to the identity of its nuclear-export sequence. *Proc Natl Acad Sci USA* **106**: 6736–6741.
21. Gonda H., Sugai M., Nambu Y. *et al.* (2003). The balance between Pax5 and Id2 activities is the key to AID gene expression. *J Exp Med* **198**: 1427–1437.
22. Gourzi P., Leonova T., Papavasiliou F.N. (2006). A role for activation-induced cytidine deaminase in the host response against a transforming retrovirus. *Immunity* **24**: 779–786.
23. Gourzi P., Leonova T., Papavasiliou F.N. (2007). Viral induction of AID is independent of the interferon and the Toll-like receptor signaling pathways but requires NF-kappaB. *J Exp Med* **204**: 259–265.
24. Greeve J., Philipsen A., Krause K. *et al.* (2003). Expression of activation-induced cytidine deaminase in human B-cell non-Hodgkin lymphomas. *Blood* **101**: 3574–3580.
25. Grimaldi C.M., Cleary J., Dagtas A.S. *et al.* (2002). Estrogen alters thresholds for B cell apoptosis and activation. *J Clin Invest* **109**: 1625–1633.
26. Grimaldi C.M., Hill L., Xu X. *et al.* (2005). Hormonal modulation of B cell development and repertoire selection. *Mol Immunol* **42**: 811–820.
27. Han J.H., Akira S., Calame K. *et al.* (2007). Class switch recombination and somatic hypermutation in early mouse B cells are mediated by B cell and Toll-like receptors. *Immunity* **27**: 64–75.
28. Hauser J., Sveshnikova N., Wallenius A. *et al.* (2008). B-cell receptor activation inhibits AID expression through calmodulin inhibition of E-proteins. *Proc Natl Acad Sci USA* **105**: 1267–1272.
29. He L., Hannon G.J. (2004). MicroRNAs: small RNAs with a big role in gene regulation. *Nat Rev Genet* **5**: 522–531.

30. Iacobucci I., Lonetti A., Messa F. *et al.* (2010). Different isoforms of the B-cell mutator activation-induced cytidine deaminase are aberrantly expressed in BCR-ABL1-positive acute lymphoblastic leukemia patients. *Leukemia* **24**: 66–73.
31. Ito S., Nagaoka H., Shinkura R. *et al.* (2004). Activation-induced cytidine deaminase shuttles between nucleus and cytoplasm like apolipoprotein B mRNA editing catalytic polypeptide 1. *Proc Natl Acad Sci USA* **101**: 1975–1980.
32. Jabara H.H., Chaudhuri J., Dutt S. *et al.* (2008). B-cell receptor cross-linking delays activation-induced cytidine deaminase induction and inhibits class-switch recombination to IgE. *J Allergy Clin Immunol* **121**: 191–196.
33. Jiang C., Foley J., Clayton N. *et al.* (2007). Abrogation of lupus nephritis in activation-induced deaminase-deficient MRL/lpr mice. *J Immunol* **178**: 7422–7431.
34. Jiang C., Zhao M.L., Diaz M. (2009). Activation-induced deaminase heterozygous MRL/lpr mice are delayed in the production of high-affinity pathogenic antibodies and in the development of lupus nephritis. *Immunology* **126**: 102–113.
35. Kincade P.W., Medina K.L., Payne K.J. *et al.* (2000). Early B-lymphocyte precursors and their regulation by sex steroids. *Immunol Rev* **175**: 128–137.
36. Lange A., Mills R.E., Lange C.J. *et al.* (2007). Classical nuclear localization signals: definition, function, and interaction with importin alpha. *J Biol Chem* **282**: 5101–5105.
37. Lanning D., Zhu X., Zhai S.K. *et al.* (2000). Development of the antibody repertoire in rabbit: gut-associated lymphoid tissue, microbes, and selection. *Immunol Rev* **175**: 214–228.
38. Lin K.I., Angelin-Duclos C., Kuo T.C. *et al.* (2002). Blimp-1-dependent repression of Pax-5 is required for differentiation of B cells to immunoglobulin M-secreting plasma cells. *Mol Cell Biol* **22**: 4771–4780.
39. Litinskiy M.B., Nardelli B., Hilbert D.M. *et al.* (2002). DCs induce CD40-independent immunoglobulin class switching through BLyS and APRIL. *Nat Immunol* **3**: 822–829.
40. MacDuff D.A., Demorest Z.L., Harris R.S. (2009). AID can restrict L1 retrotransposition suggesting a dual role in innate and adaptive immunity. *Nucleic Acids Res* **37**: 1854–1867.
41. Machida K., Cheng K.T., Sung V.M. *et al.* (2004). Hepatitis C virus induces a mutator phenotype: enhanced mutations of immunoglobulin and protooncogenes. *Proc Natl Acad Sci USA* **101**: 4262–4267.
42. Mao C., Jiang L., Melo-Jorge M. *et al.* (2004). T cell-independent somatic hypermutation in murine B cells with an immature phenotype. *Immunity* **20**: 133–144.
43. Matsumoto Y., Marusawa H., Kinoshita K. *et al.* (2007). Helicobacter pylori infection triggers aberrant expression of activation-induced cytidine deaminase in gastric epithelium. *Nat Med* **13**: 470–476.
44. McBride K.M., Barreto V., Ramiro A.R. *et al.* (2004). Somatic hypermutation is limited by CRM1-dependent nuclear export of activation-induced deaminase. *J Exp Med* **199**: 1235–1244.

45. McBride K.M., Gazumyan A., Woo E.M. *et al.* (2008). Regulation of class switch recombination and somatic mutation by AID phosphorylation. *J Exp Med* **205**: 2585–2594.
46. McCarthy H., Wierda W.G., Barron L.L. *et al.* (2003). High expression of activation-induced cytidine deaminase (AID) and splice variants is a distinctive feature of poor-prognosis chronic lymphocytic leukemia. *Blood* **101**: 4903–4908.
47. Muramatsu M., Sankaranand V.S., Anant S. *et al.* (1999). Specific expression of activation-induced cytidine deaminase (AID), a novel member of the RNA-editing deaminase family in germinal center B cells. *J Biol Chem* **274**: 18470–18476.
48. Muto T., Okazaki I.M., Yamada S. *et al.* (2006). Negative regulation of activation-induced cytidine deaminase in B cells. *Proc Natl Acad Sci USA* **103**: 2752–2757.
49. Nera K.P., Kohonen P., Narvi E. *et al.* (2006). Loss of Pax5 promotes plasma cell differentiation. *Immunity* **24**: 283–293.
50. Ordinario E.C., Yabuki M., Larson R.P. *et al.* (2009). Temporal regulation of Ig gene diversification revealed by single-cell imaging. *J Immunol* **183**: 4545–4553.
51. Park S.R., Zan H., Pal Z. *et al.* (2009). HoxC4 binds to the promoter of the cytidine deaminase AID gene to induce AID expression, class-switch DNA recombination and somatic hypermutation. *Nat Immunol* **10**: 540–550.
52. Pasqualucci L., Guglielmino R., Houldsworth J. *et al.* (2004). Expression of the AID protein in normal and neoplastic B cells. *Blood* **104**: 3318–3325.
53. Patenaude A.M., Orthwein A., Hu Y. *et al.* (2009). Active nuclear import and cytoplasmic retention of activation-induced deaminase. *Nat Struct Mol Biol* **16**: 517–527.
54. Pauklin S., Petersen-Mahrt S.K. (2009). Progesterone inhibits activation-induced deaminase by binding to the promoter. *J Immunol* **183**: 1238–1244.
55. Pauklin S., Sernandez I.V., Bachmann G. *et al.* (2009). Estrogen directly activates AID transcription and function. *J Exp Med* **206**: 99–111.
56. Petersen-Mahrt S.K., Coker H.A., Pauklin S. (2009). DNA deaminases: AIDing hormones in immunity and cancer. *J Mol Med* **87**: 893–897.
57. Rada C., Jarvis J.M., Milstein C. (2002). AID-GFP chimeric protein increases hypermutation of Ig genes with no evidence of nuclear localization. *Proc Natl Acad Sci USA* **99**: 7003–7008.
58. Ramiro A.R., Jankovic M., Callen E. *et al.* (2006). Role of genomic instability and p53 in AID-induced c-myc-Igh translocations. *Nature* **440**: 105–109.
59. Reich N.C., Liu L. (2006). Tracking STAT nuclear traffic. *Nat Rev Immunol* **6**: 602–612.
60. Reynaud C.A., Anquez V., Grimal H. *et al.* (1987). A hyperconversion mechanism generates the chicken light chain preimmune repertoire. *Cell* **48**: 379–388.
61. Reynaud C.A., Dahan A., Anquez V. *et al.* (1989). Somatic hyperconversion diversifies the single Vh gene of the chicken with a high incidence in the D region. *Cell* **59**: 171–183.
62. Reynaud C.A., Garcia C., Hein W.R. *et al.* (1995). Hypermutation generating the sheep immunoglobulin repertoire is an antigen-independent process. *Cell* **80**: 115–125.

63. Robbiani D.F., Bunting S., Feldhahn N. *et al.* (2009). AID produces DNA double-strand breaks in non-Ig genes and mature B cell lymphomas with reciprocal chromosome translocations. *Mol Cell* **36**: 631–641.
64. Rodriguez A., Vigorito E., Clare S. *et al.* (2007). Requirement of bic/microRNA-155 for normal immune function. *Science* **316**: 608–611.
65. Rogozin I.B., Iyer L.M., Liang L. *et al.* (2007). Evolution and diversification of lamprey antigen receptors: evidence for involvement of an AID-APOBEC family cytosine deaminase. *Nat Immunol* **8**: 647–656.
66. Sayegh C.E., Quong M.W., Agata Y. *et al.* (2003). E-proteins directly regulate expression of activation-induced deaminase in mature B cells. *Nat Immunol* **4**: 586–593.
67. Sernandez I.V., De Yebenes V.G., Dorsett Y. *et al.* (2008). Haploinsufficiency of activation-induced deaminase for antibody diversification and chromosome translocations both in vitro and in vivo. *PLoS One* **3**: e3927.
68. Shinkura R., Ito S., Begum N.A. *et al.* (2004). Separate domains of AID are required for somatic hypermutation and class-switch recombination. *Nat Immunol* **5**: 707–712.
69. Snapper C.M., Zelazowski P., Rosas F.R. *et al.* (1996). B cells from p50/NF-kappa B knockout mice have selective defects in proliferation, differentiation, germ-line CH transcription, and Ig class switching. *J Immunol* **156**: 183–191.
70. Sugai M., Gonda H., Nambu Y. *et al.* (2004). Role of Id proteins in B lymphocyte activation: new insights from knockout mouse studies. *J Mol Med* **82**: 592–599.
71. Ta V.T., Nagaoka H., Catalan N. *et al.* (2003). AID mutant analyses indicate requirement for class-switch-specific cofactors. *Nat Immunol* **4**: 843–848.
72. Takizawa M., Tolarova H., Li Z. *et al.* (2008). AID expression levels determine the extent of cMyc oncogenic translocations and the incidence of B cell tumor development. *J Exp Med* **205**: 1949–1957.
73. Teng G., Hakimpour P., Landgraf P. *et al.* (2008). MicroRNA-155 is a negative regulator of activation-induced cytidine deaminase. *Immunity* **28**: 621–629.
74. Thai T.H., Calado D.P., Casola S. *et al.* (2007). Regulation of the germinal center response by microRNA-155. *Science* **316**: 604–608.
75. van Maldegem F., Scheeren F.A., Aarti Jibodh R. *et al.* (2009). AID splice variants lack deaminase activity. *Blood* **113**: 1862–1864; author reply 1864.
76. Vigorito E., Perks K.L., Abreu-Goodger C. *et al.* (2007). microRNA-155 regulates the generation of immunoglobulin class-switched plasma cells. *Immunity* **27**: 847–859.
77. Withers D.R., Davison T.F., Young J.R. (2005). Developmentally programmed expression of AID in chicken B cells. *Dev Comp Immunol* **29**: 651–662.
78. Wu X., Darce J.R., Chang S.K. *et al.* (2008a). Alternative splicing regulates activation-induced cytidine deaminase (AID): implications for suppression of AID mutagenic activity in normal and malignant B cells. *Blood* **112**: 4675–4682.
79. Wu Y., Sukumar S., El Shikh M.E. *et al.* (2008b). Immune complex-bearing follicular dendritic cells deliver a late antigenic signal that promotes somatic hypermutation. *J Immunol* **180**: 281–290.

80. Xiao C., Rajewsky K. (2009). MicroRNA control in the immune system: basic principles. *Cell* **136**: 26–36.
81. Yadav A., Olaru A., Saltis M. *et al.* (2006). Identification of a ubiquitously active promoter of the murine activation-induced cytidine deaminase (AICDA) gene. *Mol Immunol* **43**: 529–541.
82. Yang G., Obiakor H., Sinha R.K. *et al.* (2005). Activation-induced deaminase cloning, localization, and protein extraction from young VH-mutant rabbit appendix. *Proc Natl Acad Sci USA* **102**: 17083–17088.
83. Zhou C., Saxon A., Zhang K. (2003). Human activation-induced cytidine deaminase is induced by IL-4 and negatively regulated by CD45: implication of CD45 as a Janus kinase phosphatase in antibody diversification. *J Immunol* **170**: 1887–1893.

Chapter 8

AID in Immunodeficiency and Cancer

Katharina Willmann,[1] Sven Kracker,[2] Maria T. Simon,[1] Don-Marc Franchini,[1] Pauline Gardes,[2] Anne Durandy,[2] and Svend K. Petersen-Mahrt[1]

[1]DNA Editing Lab, Clare Hall Laboratories, London Research Institute
Cancer Research UK
South Mimms, EN6 3LD, UK
E-mail: svend.petersen-mahrt@cancer.org.uk

[2]INSERM, U768, Hôpital Necker-Enfants Malades
Université Paris Descartes, Faculté de Médecine Paris V- René Descartes
Paris, F-75005 France
E-mail: anne.durandy@inserm.fr

The impact of the discovery of AID, with its associated function as a DNA deaminase, has been seen throughout diverse fields of studies. These include basic biochemistry, immunology, genetics and even epigenetics. Importantly, understanding the function, regulation and mechanism of AID's importance is also seen to have a direct impact on patient care. Broadly speaking, there are those patients that have an immune affliction due to mutations or even complete absence of AID (discussed in the first part of this chapter), and those patients that suffer/ed from unregulated expression of AID, which can manifest itself as cancer (discussed in the second half of the chapter). Chapter 9 of this book will look at the effect of unregulated AID expression during an immune response and how this can lead to autoimmunity. Although basic research has provided us with a large number of insights into AIDs function and oncogenic potential, it has been the precise genomic analysis of immune-deficient patients that has provided some of the most important advances in understanding its molecular mechanisms. This chapter tries to combine the analysis of immune-deficient patients as well as those of cancer aetiology as it directly pertains to AID.

8.1 AID and Immunodeficiencies

Studies of inherited immunodeficiencies have greatly contributed to a better knowledge of the normal processes of lymphocyte development, maturation and function. Among them, the recent delineation of the different class switch recombination deficiencies (CSR-D, previously named hyper-IgM or HIGM), characterized by defective antibody diversification, has provided insights into the mechanisms underlying the final step of B cell maturation. Observations of normal or high serum IgM levels, but markedly reduced serum levels of the other isotypes in these patients, strongly suggest a defect in CSR. Most patients also display impaired somatic hyper mutation (SHM). The initially described CSR-D was the X-linked form (OMIN:308230) related to mutations in the gene coding for a T cell-activation molecule, the CD40-Ligand (X67878; Armitage *et al.*, 1993; Aruffo *et al.*, 1993; Korthauer *et al.*, 1993). More recently, mutations in its counterpart, the CD40 molecule (X60592), have been reported as responsible for a very rare autosomal recessive (AR) CSR-D with a similar phenotype (OMIN:606843; Ferrari *et al.*, 2001). Other CSR-D are caused by an intrinsic defect likely to be affecting the CSR machinery in B cells.

8.1.1 *Autosomal recessive CSR-D caused by bi-allelic Aicda mutations*

The most frequent cause of CSR-D is mutations in the *Aicda* gene (AB 040430), encoding the activation-induced cytidine deaminase molecule (OMIM:605258; AID; Revy *et al.*, 2000). From this publication, four other reports describing AID deficiency (MIM #605258) have been published in the literature (Durandy *et al.*, 2006; Lee *et al.*, 2005; Minegishi *et al.*, 2000; Quartier *et al.*, 2004).

8.1.1.1 *Clinical phenotype*

In our cohort of 64 patients the clinical phenotype is very similar. The onset of symptoms occurs during early childhood (mean age, 5 years) and is characterized by a high susceptibility to recurrent bacterial infections: more than half of patients (58%) had respiratory tract infections, with 14% reporting bronchiectasis. Gastrointestinal infections were observed in 27% of cases, sometimes related to persistent Giardia infections. Such infections may result in failure to thrive. Infections of the central nervous system (e.g., meningitis) have been reported in 25% of AID-deficient patients, often associated with inadequate Ig substitution. One case of herpes virus encephalitis and one of poliomyelitis has been described. Two adult patients died prematurely, one of pulmonary haemorrhage at 47 years of age, the other of septicemia at age 63. One patient developed an Erwing sarcoma, and another one a B cell lymphoma.

A striking lymphoid hyperplasia is present in the majority (75%) of patients, affecting predominantly the cervical lymph nodes and tonsils. In one case, mesenteric lymph node hyperplasia resulted in intestinal obstruction. Hepatosplenomegaly has been reported in 10% of the patient cohort. Other manifestations include arthritis (12%) and autoimmune manifestations (hemolytic anaemia, thrombopenia and autoimmune hepatitis) in 21%, with autoantibodies of the IgM isotypes detectable in some cases. Systematic lupus erythematous, diabetes mellitus and Crohn's disease have been reported in one patient each.

8.1.1.2 *Laboratory findings*

All patients had normal or elevated IgM at the time of diagnosis and markedly undetectable serum levels of IgG and IgA. In agreement with the aberrant serum immunoglobulin levels, antigen-specific (vaccine or infectious agents) antibodies of the IgG isotype were not detectable. When analyzed, IgM iso-hemagglutinins and antipolysaccharide IgM antibodies were present. IgM serum levels often diminish after Ig substitution, a finding suggesting that increased IgM reflects chronic antigenic stimulation rather than a direct effect of the AID deficiency.

Numbers of peripheral blood T cells (CD3$^+$) and T cell subsets (CD4$^+$ and CD8$^+$), as well as *in vitro* T cell proliferation to mitogens and antigens are normal. Peripheral blood B cells (CD19$^+$) are normal in number and normally express CD40 molecules. All CD19$^+$ B cells express or co-express sIgM and sIgD, in contrast to age-matched controls, who have a population of CD19$^+$ B cells that do not express sIgM nor sIgD and therefore most likely to have already undergone CSR.

Soluble CD40-L (sCD40L)-induced B cell proliferation *in vitro* is normal. However, *in vitro* activation of B lymphocytes by sCD40L in combination with appropriate cytokines (IL-4 or IL-10), which induces, respectively, IgE or IgA production in controls and CD40L-deficient patients (Durandy *et al.*, 1993), is ineffective in AID-deficient patients. Under the same culture conditions, switch (S) region transcription is normally induced, whereas DNA double-strand breaks are not detected in Sμ regions, providing evidence that the CSR defect is located downstream from transcription and upstream from DNA cleavage (Catalan *et al.*, 2003).

Similarly to healthy age-matched controls, a fraction (20–50%) of patients' B cells express the CD27 marker, which has been described as a marker of mutated memory B cells, even in the IgM$^+$/IgD$^+$ B cell compartment (Klein *et al.*, 1998). However, the frequency of SHM in the IgM V region on CD19$^+$CD27$^+$ B cells from patients with bi-allelic AICDA mutations is null or dramatically reduced as compared to age-matched controls. Thus, AID deficiency leads to lack of both CSR and SHM (Revy *et al.*, 2000).

Most of the patients exhibit enlarged secondary lymphoid organs, which may require surgical resection (tonsils) or biopsy (cervical lymph nodes); histological evaluation shows marked follicular hyperplasia. Germinal centers are giant, being two- to more than 10-times larger than those from control reactive lymph nodes. The mantle zone and interfollicular areas appear thin. The giant germinal centers contain a normal follicular dendritic cell network and B cells that are PNA$^+$, CD38$^+$, CD23$^+$, CD83$^+$, CD95$^+$, CD40$^+$, IgM$^+$, Bcl2$^+$ and Ki67$^+$. Strikingly, many germinal center B cells from patients express sIgD, in contrast to normal germinal-center B cells, in which IgD$^+$ B cells are rarely found. Occasional CD27$^+$ B cells as well as IgM- and IgD-expressing plasma cells are found in germinal centers and T cell areas; however, neither IgG- nor IgA-expressing plasma cells could be observed. In the few cases studied, AID expression was not detectable in germinal center B cells from patients with bi-allelic *AICDA* mutations by immunohistochemistry.

8.1.1.3 *Molecular basis of autosomal recessive AID deficiency*

Our series includes 64 patients from 45 families, 27 of which were known to be consanguineous. We found 34 different gene alterations, most often as homozygous defects (34 families), less frequently as compound heterozygous mutations (12 families). Mutations were scattered all along the gene, with no peculiar hotspot. They included missense mutations (20 families), nonsense mutations (five families) and small deletions (in frame: two families; out-of-frame: one family). In addition, deletions of the entire coding region (two families) or splice-site mutations (five families) leading to either a longer RNA transcript (one family) or to frameshift and premature stop codon (five families) were observed (Fig. 8.1). Since the same mutations were found in several unrelated families with the same ethnic origin, analysis of flanking polymorphic markers was performed, indicating a common ancestral origin of the mutation (Revy *et al.*, 2000).

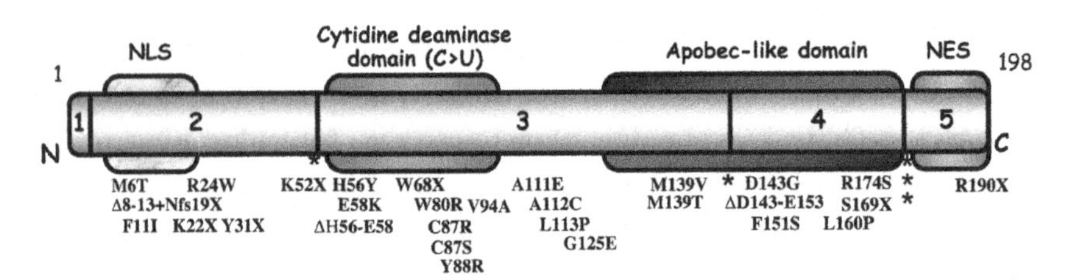

Figure 8.1. ***Aicda* mutations observed in 64 patients.** * splice defect; two families present with deletion of the whole genetic region as assessed by Southern blot analysis.

Three different missense mutations were located in the N-terminus of AID in 10 patients, immediately upstream from the putative nuclear localisation signal (NLS, M6T) or within the NLS (F11I, R24W, Fig. 8.1). All 10 patients presented with a typical CSR defect as shown by their Ig levels (SHM was also lacking, in the cases where such analyses were performed). Conversely, artificial *Aicda* mutants with mutations located in the NLS have been shown to possess, as expected, normal cytidine deaminase activity after transfection in *E. coli*, and strikingly, normal CSR activity when transfected into fibroblasts (Shinkura *et al.*, 2003). In contrast, such mutants showed defective SHM induction in model systems (Shinkura *et al.*, 2003). This study raised the hypothesis that AID could play a role in SHM by means of interaction of its N-terminal part with specific SHM-cofactors. However, this discrepancy between normal CSR and defective SHM was not observed in the patients we could study, even when the expression level of the mutated AID (R24W) was normal. This observation might suggest that recruitment of cofactors would not be identical in humans and mice, or alternatively, that N-terminal AID mutants are unable to properly shuttle into the nucleus. One can, however, not exclude the existence of a new immunodeficiency characterized by defective antibody affinity maturation but normal CSR, and caused by AID mutations located at the N-terminus, provided that AID protein is expressed and traffics normally between cytoplasm and nucleus upon activation.

Conversely, in three cases, although CSR was dramatically impaired, a normal frequency and nucleotide substitution pattern of SHM was observed. The giant germinal centers usually observed in secondary lymphoid organs in other AID-deficient patients were not detected in a lymph node biopsy performed in one of these cases. Strikingly, all three patients carry mutations located in the C-terminal part of *Aicda* (Fig. 8.1). In one, a homozygous mutation in the splice acceptor site of intron 4 leads to the in-frame insertion of 31 amino acids. The other patients exhibit compound mutations: first, a missense mutation in the APOBEC-1-like domain (R112C) and a mutation in the splice donor site of intron 4, leading to a 26 amino-acid frameshift replacement of the C-terminus by 28 amino acids leading to the nuclear excision signal (NES) sequence loss; and second, a mutation in the cytidine deaminase domain (P83S) and a splice donor site mutation of intron 4 resulting in deletion of exon 4 and frameshift insertion of four amino acids also excluding the NES. The phenotype of these patients demonstrates that AID plays a major role in CSR not only through its cytidine deaminase activity but also likely to be by recruitment of specific cofactors. Although no data are available in humans, *in vitro* experiments indicate that similar mutants, inefficient in CSR, do normally induce mutations and DNA double-strand breaks in Sμ regions, suggesting rather a defect in DNA repair than abnormal targeting of AID to the S regions (Barreto *et al.*, 2003; Doi *et al.*, 2009). Although several AID cofactors have been recently described (Teng *et al.*, 2008; Conticello *et al.*, 2008; Vuong *et al.*, 2009), none have been shown to exert a differential activity on CSR and SHM. However, several pieces of evidence strongly suggest that the C-terminal part of AID interacts with a still unknown cytoplasmic CSR cofactor (Geisberger *et al.*, 2009; Doi *et al.*, 2009). Interestingly, some patients have been shown to suffer from a defective CSR associated with a DNA repair abnormality (Peron *et al.*, 2007). In these patients, *Aicda* sequence is normal and AID is expressed. Although in these patients, SHM pattern is found slightly different from controls, one can imagine that these patients suffer from a defect in a putative CSR AID cofactor, which is also involved in DNA repair.

8.1.2 *Autosomal dominant CSR-D caused by mono-allelic Aicda mutations*

In addition, in 11 patients from five families we have observed a CSR-D transmitted as an autosomal dominant disease and characterized by variably increased serum IgM levels and IgG and IgA defects. All patients have the same heterozygous nonsense mutation (R190X) resulting in the NES loss (Imai *et al.*, 2005). Both normal and mutated allele *Aicda* transcripts were equally expressed in activated B cells as judged by semi-quantitative RT-PCR amplification. *In vitro* studies showed a complete defect in CSR, localized downstream from the DSB in Sμ regions while SHM frequency was found to be normal, except in one patient. The variable immunodeficiency found in R190X heterozygous patients is not caused by haploinsufficiency, considering the normal Ig levels observed in subjects carrying heterozygous mutations were located elsewhere in the *Aicda* gene (NLS, cytidine deaminase and APOBEC-1-like domains). Two non-mutually exclusive hypotheses can be proposed to account for the dominant negative effect exerted by R190X allele. First, AID forms homomultimeric complexes *in vitro* (Ta *et al.*, 2003; Patenaude *et al.*, 2009), even in the absence of its C-terminal part, as shown by normal *in vitro* multimerization of artificial mutants truncated for the last 10 amino acids (Barreto *et al.*, 2003). The R190X mutated allele can therefore be incorporated in the AID multimeric complexes, impairing their activity through a dominant effect. Second, the truncation affects the NES domain of the protein, leading to a lack of cytoplasmic retention and nuclear accumulation of the mutant allele, which could interfere with normal AID trafficking and function. While wild-type AID is only detected in the cytoplasm of activated B cells, being able to shuttle into the nucleus for a short period of time, the R190X mutated allele was found in the nucleus after overexpression in AID-deficient B cells (Ito *et al.*, 2004). However, the observation that two different splice site mutations leading to insertion and truncation of the NES (see above) are responsible for an autosomal recessive disease, suggests that a conformational change rather than the loss of NES is responsible for the dominant negative exerted by the R190X allele.

The ongoing delineation of inherited CSR-D is shedding new light on the process of physiological antibody maturation in humans. Studies of patients harboring different mutations in the *Aicda* gene, together with the data from artificial mutants reported in the literature, revealed unexpected functions of AID, a master molecule for antibody maturation: AID appears not only as a DNA-editing enzyme but also as a docking protein recruiting cofactors in a multimolecular complex.

8.2 AID and Cancer

To date, almost all cancers that have been studied contain alterations in their genome when compared to their 'normal' counterparts. These alterations usually manifest themselves as mutations or translocations of key proteins or regulatory sites within the DNA, but on occasion may also be of epigenetic origin (alteration in the make-up of the chromatin structure rather than the DNA *per se*). Surprisingly, considering the prevalence and diversity of tumor origin, direct links between environmental carcinogens and DNA mutation in human tumors has been limited (Weinberg, 2007). Over the past 30 years, a number of mutagens have been identified that do not seem to cause cancer (as assessed from epidemiological evidence); in contrast, non-mutagens can induce oncogenesis, with more than 40% of tumor-inducing agents (in rat models) not causing DNA mutations in mutagenesis tests (Weinberg, 2007).

A review on AID in oncology will be different from that of most other genes, including oncogenes and tumor suppressors, as the function of DNA deaminases is to induce DNA damage and therefore their oncogenic potential is naturally high. Therefore, DNA deaminases could be classified both as oncogenes as well as a chemical mutagen.

8.2.1 *AID is a mutagen*

Although after initial isolation and characterization AID was classified as an RNA-editing enzyme (predominantly based on its homology to APOBEC1), it is now well established that AID is the vertebrate evolutionary founding member of a class of enzymes that deaminate cytosine residues in single-stranded DNA (Conticello *et al.*, 2005). Cytosine deaminase activity was identified and inferred by *in vitro*, in *E. coli*, and *in vivo* experimentation. It is therefore likely, that in any tissue where expression of a DNA deaminase (especially AID) can be found, DNA alterations will occur. It is less clear how efficiently the cellular milieu regulates the activity of DNA deaminases and how efficiently the cell deals with resolving the damage. Intriguingly, the majority of single-point oncogenic mutations (Greenman *et al.*, 2007), as well as the majority of single nucleotide polymorphisms, in humans are C (or G) transitions (Sachidanandam *et al.*, 2001).

8.2.2 *AID is a carcinogen*

As mentioned above, not every mutagen is a carcinogen, yet transgenic overexpression of AID has demonstrated tumor formation in various tissues (Endo *et al.*, 2007; Matsumoto *et al.*, 2007; Okazaki *et al.*, 2003b). Interestingly, mice seem to be normal initially, and only in the late stages of life were a significant number of tumor formations detected. Therefore, it is plausible that in these mice the effects of AID on tumor formation are secondary rather than primary. In favour of this hypothesis is the estimate that normal human cells have a large number of lesions in their genome at any given time (20,000 single-stranded breaks, 10,000 depurinations, 5,000 alkylating lesions, 2,000 oxidative lesions, 600 deaminations and 50 sister chromatid crosses [Lindahl, 1993]). If most of these are efficiently repaired, a 'few more' AID-induced deaminations are not sufficient to drive oncogenesis.

A number of years ago it was established that AID is a prerequisite for the formation of the oncogenic translocation c-Myc/IgH (Pasqualucci *et al.*, 2004; Ramiro *et al.*, 2004; Robbiani *et al.*, 2008). During class switch recombination, AID could have been mis-targeted to the c-*myc* proto-oncogene loci, and subsequent aberrant DNA repair has led to a translocation between the two loci. This placed the c-*myc* expression under the control of the immunoglobulin loci (Ig). The tumor potential of this translocation had already been identified in Burkitt lymphomas as early as the 1980s (Rabbitts *et al.*, 1984). Recent work showed that AID is required for the targeting and induction of, DNA instability in the c-*myc* locus (Robbiani *et al.*, 2008), and demonstrated AID's role as a tumor-inducing agent (Robbiani *et al.*, 2009).

Aside from Burkitt lymphomas, other leukemias have been analyzed for a requirement of AID expression. In a recent study, Klemm *et al.* were able to demonstrate that AID could also play a role in chronic myeloid leukemia (Klemm *et al.*, 2009). The authors were able to demonstrate that the clinical progression, as well as drug resistance in the BCR-ABL1 gene, were in part due to the continuous expression of AID. This indicated that AID can also play an important role in tumor progression rather than just initiation.

8.2.3 *Cancer markers and AID*

One of the hallmarks of neoplastic cells is that they no longer express a 'normal' set of genes. In the past decade, tumor markers have become valuable diagnostic tools in understanding cancer progression, as well as providing information for patient care. These 'markers' though, are not always involved in the actual molecular biology of tumor formation/progression, but can represent a bystander effect. Unlike transformed oncogenes (i.e. proto-oncogenes that have altered protein structure, sequence, modification or expression) DNA deaminases can induce their oncogenic potential (mutation of a cytosine) as part of their 'normal' activity. They can be activated by stress, hormones and other environmental signals, and return to 'un-stressed' levels subsequently – even in a transformed cell. Regardless, with DNA deaminases being both mutagen and oncogene, a number of cancers (especially of B cell origin) have been analysed for their expression of AID and potential prognostic association.

There are more than 20 publications (to date) that refer to AID as a potential marker in various B cell cancers (Babbage *et al.*, 2004; Dijkman *et al.*, 2006; Engels *et al.*, 2008; Feldhahn *et al.*, 2007; Forconi *et al.*, 2004; Greeve *et al.*, 2003; Greiner *et al.*, 2005; Guikema *et al.*, 2005; Hardianti *et al.*, 2004; Hardianti *et al.*, 2005; Klemm *et al.*, 2009; Lossos *et al.*, 2004; McCann *et al.*, 2009; Moldenhauer *et al.*, 2006; Montesinos-Rongen *et al.*, 2005; Pasqualucci *et al.*, 2004; Popov *et al.*, 2007; Reiniger *et al.*, 2006; Ruminy *et al.*, 2008; Smit *et al.*, 2003; Willenbrock *et al.*, 2009). It is beyond the scope of this review (as well as the authors) to make a proper judgement on this topic, considering the potential confusion arising from the above mentioned ability for AID to recede back to 'normal' after its oncogenic potential has unfolded. Although AID is not a reliable marker for tumor diagnosis, its expression in a subset of various lymphomas can provide new insight into the origin of tumors, and that in turn can assist in diagnosis and treatment (Feldhahn *et al.*, 2007; Hardianti *et al.*, 2005; Lossos *et al.*, 2004). Furthermore, AID's involvement in drug escape (see above) will provide new emphasis on AID as a pharmacological target. Analysis of AID expression should also help in determining the mechanisms of B cell lymphomas. Part of the disparity in identifying AID as a proper marker in the progression of B cell pathologies may stem from the observation that the 'pluripotency activator' Pax5 is active in B cells (Cobaleda and Busslinger, 2008) and may provide a mechanism by which 'normal' developmental progression is altered. Pax5 (aside from altering AID expression [see below]) can induce, under certain circumstances, a pluripotency state in B cells (Cobaleda and Busslinger, 2008) – in effect reversing B cell development. Thereby, the notion "...that malignant B cells seem to be 'frozen' at a particular differentiation stage, which reflects their origin." [(Kuppers, 2005) and references within] may no longer be true.

8.2.4 *AID regulation and cancer correlation*

Evidence from a number of different laboratories has shown that, in B cells, there is a strong correlation between AID concentration and translocation potential (Dorsett *et al.*, 2008; Teng *et al.*, 2008; Takizawa *et al.*, 2008). Therefore, any events leading to an alteration in AID expression, modification, localisation or activity has the potential to be oncogenic. Figure 8.2 depicts a schematic (and Table 8.1 tabulating) of a

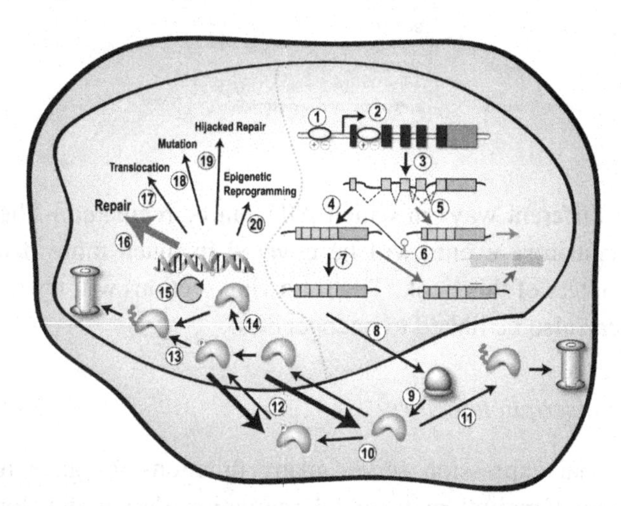

Figure 8.2. **Schematic of AID expression and possible regulatory points within a cell.** Numbers next to arrows indicate biogenesis pathway and corresponds to numbers in Table 8.1.

Table 8.1. **Possible events regulating AID biogenesis and function**

	Event	Comment	Chapter
1	Enhancer	not determined yet	7
2	Promoter	NFkB, ERa, Pax5, Id2, HoxC4, STAT6, Sp1	7
3	Transcription	EBV, HCV, Abl-MLV, H. pylori, estrogen	7
4	Splicing	not determined yet	
5	Alt Splicing	possible AID inactivation	
6	miRNA	miRNA155, miRNA181b	7
7	mRNA processing	not determined yet	
8	mRNA transport	not determined yet	
9	Translation	not determined yet	
10	Phosphorylation	PKA, PKC	4, 7
11	Ubiquitination	not determined yet	7
12	Nuclear Shuttling	importin alpha and beta	4, 7
13	Ubiquitination	nuclear AID degradation	
14	Targeting	off-target Bcl-6, c-Myc, Pax5, FAS, RHO/TTF, PIM1, BCR-ABL	2, 3, 4
15	Processive Distributive	AID vs. APOBEC3G	
16	normal DNA repair	DNA-PK	5, 6
17	Translocation	c-Myc/IgH	5
18	Mutations	BCR-ABL	6
19	Hijacked Repair	MMR, BER, TLS	5, 6
20	Demethylation	epigenetics and demethylation	

number of different ways in which AID can be regulated. The majority of these regulatory events will be covered in much more detail in the various chapters of this book. The following sections will try to highlight how they can also be linked to oncogenesis.

8.2.4.1 *Transcription factors*

Transcriptional expression serves many functions for gene regulation, ranging from temporal and spatial regulation during development, to cell-cycle-specific activation and environmental stimulus responses. A number of specific key transcription factors have been identified that modulate AID gene expression, including NF-κB, STAT6, HoxC4, Sp1, Pax5, E47, Id2, Id3 and ERα (Dedeoglu *et al.*, 2004; Gonda *et al.*, 2003; Park *et al.*, 2009; Pauklin *et al.*, 2009; Sayegh *et al.*, 2003; Yadav *et al.*, 2006). Because of their key function, some of these factors have been implicated as oncogenes or tumor suppressors.

The nuclear factor-κB (NF-κB) is a key player in the development of various tissues, and plays a controlling role in both innate and adaptive immunity. The authors feel that they would do a disservice to the large number of key contributors in the field of NF-κB, if they were to list a limited number of references here. Rather, we ask the reader to look at text- books in immunology, development, biochemistry and cancer to appreciate the importance of AID regulation by NF-κB, but suffice to say that constitutive activation of NF-κB leads to various lymphomas.

Activated STAT6 (Bruns and Kaplan, 2006) has been detected in different forms of cancer including primary mediastinal large B cell lymphomas (PMBL) and Hodgkin lymphomas (Ni *et al.*, 2002; Skinnider *et al.*, 2002). Recent analysis of primary lymphoma from patients has revealed mutations in the DNA-binding domain of STAT6 and an increase DNA-binding activity of STAT6 (Ritz *et al.*, 2009).

Although the role of HoxC4 in cancer progression was established over a decade ago, little else is known about its oncogenicity aside from its predominant expression in lymphoid cells and overexpression being detected in mantle cell lymphoma, Burkitt's lymphoma and B cell chronic lymphocytic leukemia (Bijl *et al.*, 1996; Bijl *et al.*, 1997; Miller *et al.*, 2003). The recent publication of how HoxC4 can bind to the AID promoter has highlighted that CD40 signaling, LPS and cytokines, but also hormones (unpublished data), induce HoxC4 expression (Park *et al.*, 2009).

The regulation of AID by Pax5 is of particular interest in cancer, as Pax5 can control the AID locus (Gonda *et al.*, 2003), while at the same time Pax5 can be targeted by AID for mutations (Pasqualucci *et al.*, 2001). SHM of Pax5 has been found in diffuse large B cell lymphoma (DLBCL; Pasqualucci *et al.*, 2001), while CSR between IgH and Pax5 – the t(9;14) translocation (Busslinger *et al.*, 1996; Iida *et al.*, 1996;

Morrison *et al.*, 1998) – is mainly associated with aggressive B cell non-Hodgkin lymphoma (Poppe *et al.*, 2005). Aside from its important role as a B cell plasticity factor (Cobaleda and Busslinger, 2008), Pax5 can also play a role in cancer progression outside the B cell lineage, where reactivation of Pax5 in myeloid leukemia-induced AID as well as malignant progression (Klemm *et al.*, 2009). Id2, a member of the Id protein family controlling cell cycle and cell fate, is the transcription factor counteracting the Pax5 activity. Members of this family have been implicated in other cancers (Benezra *et al.*, 2001; Iavarone and Lasorella, 2004; Lasorella *et al.*, 2001).

8.2.4.2 *Extracellular stimuli*

Environmental factors that can alter AID expression as well as being oncogenic include bacteria/viruses and hormones (see Chapter 7). Interestingly, viral induction can have both oncogenic and anti-oncogenic potential. EBV infection can activate AID (Epeldegui *et al.*, 2007), but at the same time EBV has long been associated with the formation of Burkitt lymphomas (Weinberg, 2007). EBV infection also induces error-prone polymerase polη expression, which, along with AID, possibly contributes to the enhancement of mutations found in BCL6 and p53 proto-oncogenes (Epeldegui *et al.*, 2007; Gil *et al.*, 2007).

Hepatitis C virus (HCV) is able to infect hepatocytes as well as B cells and causes hepatocellular carcinoma and non-Hodgkin's B cell lymphoma (Weinberg, 2007). Infection promotes AID expression in B cells isolated from peripheral blood, and analogous to EBV infection, induces error-prone polymerases (Machida *et al.*, 2004), with mutations driven by AID identified in the IgH locus, BCL6, p53 and β-catenin genes. Furthermore, AID expression has also been detected in hepatocytes after HCV infection (Kou *et al.*, 2007).

AID activation in B cells was also seen following infection with Abelson murine leukemia virus (Abl-MLV; Gourzi *et al.*, 2006). Infection of B cells can be achieved *in vitro*, while *in vivo* infections induce acute pre-B cell leukemia (Rosenberg and Papavasiliou, 2007). Importantly, AID acted as a guardian for the genome, as B cells deficient for AID appeared to be more susceptible to the infection. Furthermore, since p53 turns out to be mutated in 50% of infected cells and p53 deficiency is known to enhance AID-induced c-myc/IgH translocations (Ramiro *et al.*, 2006), it is likely that AID enhanced the leukemogenic process.

In addition to these viruses, Matsumoto and colleagues showed that *in vitro Helicobacter pylori* infection induced AID expression in epithelial gastric cells, and that AID was detected in 78% of human gastric cancer specimens (Matsumoto *et al.*, 2007). As in the case of viral infection, AID activity correlated with the incorporation of mutation in the p53 tumor suppressor gene. *Helicobacter pylori* had been discovered in the 1980s, yet the initial claim to induce gastric cancers was dismissed (Weinberg, 2007). It is still not clear what the precise molecular mechanism is that provides the now accepted role in oncogenesis, but *Helicobacter pylori's* activation of the NF-κB pathway could play an important role (Maeda *et al.*, 2002; Hamajima *et al.*, 2006).

To date, of the chemical or soluble factors that can activate AID, estrogen has been the most extensively studied compound. It is also one of the oldest known cancer-causing agents [(Pauklin *et al.*, 2009; Petersen-Mahrt *et al.*, 2009) and references within]. The ability of estrogen to bind to the estrogen receptor in the cytoplasm, its subsequent translocation to the nucleus, and its binding to the AID promoter, provides the most direct link between a non-genotoxic environmental agent and a DNA mutation system. It also provides a new insight into how gender bias immune diseases, such as systemic lupus erythematosus, could have a direct 'drugable' molecular target (Maul and Gearhart, 2009).

8.2.4.3 *mRNA biogenesis*

RNA pol II transcripts have to be processed into functional mRNAs, exported to the cytoplasm as RNPs (ribonuclear particles) and translated by ribosomes. Throughout this biogenesis a number of key events can act as regulatory steps, altering the overall expression of the protein within a cell. Currently, there is no data for alternative poly-adenylation or capping of AID mRNA. Although splicing has long been recognized as a means of gene control, alternative splicing has risen to prominence since it has been shown that more than 40% of the 30,000 genes in the human genome are alternatively spliced. More recently, *trans*-acting splicing proteins and *cis*-acting splice regulatory sequences have been shown to play an important role in certain cancer etiologies (Srebrow and Kornblihtt, 2006). Although work by different laboratories have identified splice variants of AID (Albesiano *et al.*, 2003; van Maldegem *et al.*, 2009; Wu *et al.*, 2008), it still needs to be determined what the precise function of those variants are and to what extent they may represent a true regulatory function. Even if they are only used to inactivate AID's normal function, they would provide a means of control, and lack of such 'inactivating' alternative splicing could thus be oncogenic.

There are a number of different ways how the half-life of specific mRNAs can be regulated, most of which reside within the 3′ UTR. The importance of AID's mRNA stability has been highlighted by the discovery that miRNAs affect AID expression (Dorsett *et al.*, 2008; Teng *et al.*, 2008; de Yebenes *et al.*, 2008) – for the precise mechanism see Chapter 7.

miRNAs are small single-stranded RNAs derived from precursors through processing by RNAse III enzymes Drosha and Dicer. They have been found to be important in development, cell differentiation and regulation of cell cycle and apoptosis. Mechanistically, miRNAs control gene expression function via sequence-specific base-pairing with mRNA, causing inhibition of translation or degradation of the target mRNA (Bartel, 2004; Leung and Sharp, 2006). Because of this potential ubiquitous function, miRNAs have been shown to be involved in the initiation and progression of many cancers [(Croce, 2009; Navarro *et al.*, 2009; Negrini *et al.*, 2007) and references within].

A number of miRNAs have been implicated directly in oncogenesis, including miR-155. Generated from the Bic precursor gene, it was first implicated in the induction of B cell lymphomas by Avian Leukosis Virus (ALV) in chickens, where it is thought to be associated with integration of ALV into the Bic locus (Tam *et al.*, 2002). Furthermore, miR-155 overexpression has been documented in a number of lymphoid neoplasms such as chronic lymphocytic leukemia, diffuse large B cell lymphomas, Hodgkin's lymphoma and primary mediastinal large B cell lymphoma (Costinean *et al.*, 2006; Eis *et al.*, 2005; Kluiver *et al.*, 2005). Interestingly, some lymphomas (i.e. Burkitt's lymphoma) express low bic/miRNA-155 (Kluiver *et al.*, 2006) or have a processing defect in miR-155 (Kluiver *et al.*, 2007).

The 3′ UTR of AID was found to possess a target site for miR-155 (Dorsett *et al.*, 2008; Teng *et al.*, 2008) and AID expression was found to be elevated in miR-155-deficient B cells (Vigorito *et al.*, 2007). When the miR-155:AID mRNA interaction was de-regulated, AID protein expression was persistent, due in part to a two-fold increase in AID mRNA half-life, leading to increased c-myc-IgH translocations (Dorsett *et al.*, 2008; Teng *et al.*, 2008). Not only did this work shed light on how AID regulation can lead to translocations, but it also provided the miRNA field with a novel and direct mechanism for oncogenesis.

Aside from miRNA-155, miRNA-181b can also alter AID mRNA expression (through the 3′ UTR) and with it affect translocation frequencies (de Yebenes *et al.*, 2008). Although it is not yet known if this is cause or consequence, recent work has demonstrated miRNA-181B to be altered in various cancers (Careccia *et al.*, 2009; Conti *et al.*, 2009; Nakajima *et al.*, 2006; Pallasch *et al.*, 2009; Schmitz *et al.*, 2009; Shi *et al.*, 2008).

Although some of the above pathways are listed as independent control mechanisms for AID expression, there are of course numerous instances where the combination of two or more pathways can contribute to AID's oncogenic potential. Examples of these are the interaction of the NFκB and estrogen transcription pathway (de Bosscher *et al.*, 2006; Stice and Knowlton, 2008), the SP1 transcription factor and hormone pathway (Solomon *et al.*, 2008), virally-activated cytokines and the stress-response (NF-κB) pathway [e.g. HCV, in combination with proinflammatory cytokines TGFβ, TNF-α and IL-1β (Endo *et al.*, 2007; Kou *et al.*, 2007)], or even estrogen's ability to influence miRNA processing (Yamagata *et al.*, 2009). The latter being of particular interest, as a recent report indicated that miR-155 has also been found to be differentially expressed in the serum of women with hormone-sensitive breast cancer compared to women with hormone-insensitive breast cancer (Zhu *et al.*, 2009).

8.2.4.4 *Protein production and modification*

Translation itself has not yet been shown to influence AID expression, but a number of post-translational modifications are known to do so. As seen in Chapter 4, phosphorylation of AID can play a critical role in its function and activity. Two serine/threonine kinases have been identified that phosphorylate AID – Protein Kinase A and C (PKA and PKC) – both of which have been implicated in oncogenesis.

PKA belongs to a large superfamily of proteins whose activity is regulated by cyclic AMP. PKA overexpression has been found to be associated with poor patient outcome for colorectal, breast, lung and prostate cancer (Bradbury *et al.*, 1994; Pollack *et al.*, 2009). This appears to be particularly true for metastatic cancers. Patients with inactivating mutations in the PRKARIA gene (which encodes the type 1A regulatory subunit of PKA) suffer from an autosomal dominant tumor predisposition called Carney complex (Boikos and Stratakis, 2006). PKA has also been shown to be able to alter the estrogen (see above) signalling cascade (Al-Dhaheri and Rowan, 2007), thereby providing another means of regulating AID.

PKC is one of the most extensively studied kinases in cancer, as it has been shown that a subset of this family can be directly activated by the tumor promoter phorbol ester (Weinberg, 2007). Mutations and chromosomal rearrangements in PKC isozymes are rare, however, altered expression of PKC isozymes has been observed in many cancers (Cornford *et al.*, 1999; Haughian *et al.*, 2006; Lahn *et al.*, 2004; Mischak *et al.*, 1993; Murray *et al.*, 1999; O'Brian *et al.*, 1989). PKCα levels, particularly, have been found to be increased in breast cancer patients with low/negative estrogen receptor levels (Lahn *et al.*, 2004; O'Brian *et al.*, 1989), although in other cancers this isoforms expression is downregulated [e.g. colon (Kusunoki *et al.*, 1992)]. Of note is the fact that PKCε, generally found to promote cell survival (Mischak *et al.*, 1993), has been shown to be upregulated in various types of cancers including prostate cancer (Cornford *et al.*, 1999). Furthermore, PKCε overexpression has been linked to chemotherapeutic resistance in various cell types (Ding *et al.*, 2002; Tachado *et al.*, 2002). Finally, several PKCs have been implicated in invasion and metastasis of cancer cells (Platet *et al.*, 1998; Zhang *et al.*, 2004).

Although indicated in Fig. 8.2 to take place in the cytoplasm as well, it is likely that the majority of the AID ubiquitination for proteosomal degradation takes place in the nucleus (Aoufouchi *et al.*, 2008). As with protein expression, protein degradation plays a pivotal role in regulation of key proteins. The pleiotropic usage of ubiquitin has two predominant functions, either as a signaling molecule that would be equivalent to phosphorylation, or as a means to initiate protein turnover. To date, ubiquitin modifications of AID have only been identified for degradation (Aoufouchi *et al.*, 2008), but it is likely that there will be instances where ubiquitin or ubiquitin-like modifications will serve other functions. Because of its widespread usage in all aspects of cellular function, it is not surprising that ubiquitin modifications play an important role in cancer (Hoeller *et al.*, 2006) and have become a sought after drug target (Hoeller and Dikic, 2009). Aside from targeting key tumor suppressor proteins, the ubiquitin ligases have been identified as oncogenes and when misregulated can induce cancer. For instance, excessive SKP2 in malignant melanoma and lymphoma can target the tumor suppressor p27/KIP (Katagiri *et al.*, 2006; Lim *et al.*, 2002). If the yet to be identified AID ubiquitin ligase is inactive, excessive AID may lead to oncogenesis.

8.2.4.5 *Subcellular localization and targeting*

It is still to be determined which of the above mentioned modifications of AID will have a direct impact, and at which stage, on the overall activity of AID, but it is clear that subcellular localization and targeting will play a very important role. Already from the earliest overexpression data, it became clear that AID was predominantly found in the cytoplasm – away from its substrate DNA (Rada *et al.*, 2002). It is now accepted that this is part of the cellular regulation to control the function of AID. The molecular partners and mechanism of how AID is transported into and out of the nucleus is still to be determined (see Chapters 4 and 7), but recent work has begun to reveal a complex system of localisation and retention signals in AID itself (Patenaude *et al.*, 2009), as well as some interacting proteins. One such protein is the developmentally-important protein importin α1. This interaction would link AID to a highly regulated subcellular distribution pathway. Although importin α mutations have not been shown to be part of a transformed phenotype, importin α has been shown to be an important marker for breast cancer prognosis (Gluz *et al.*, 2008). The nuclear export of AID seems to depend on the CRM1 export pathway (Brar *et al.*, 2004; Ito *et al.*, 2004; McBride *et al.*, 2004). This pathway may play an important role in cancer, as it transports tumor suppressors such as p53, FOXO and p21/p27 into the cytoplasm (Turner and Sullivan, 2008), and has recently been identified as a potential drug target (Turner and Sullivan, 2008; Mutka *et al.*, 2009).

AID activity is most prominently observed at the Ig locus, and even there it is restricted to a few defined regions (see Chapters 2 and 3), yet a number of off-target activities (SHM) have been reported even prior to AID isolation. Aberrant CSR can trigger c-myc translocations (Rabbitts *et al.*, 1984; Ramiro *et al.*, 2004) and aberrant SHM in genes such as BCL6, CD95/FAS, RHO/TTF, PAX-5 and PIM1 (Pasqualucci *et al.*, 2001; Pasqualucci *et al.*, 1998; Muschen *et al.*, 2000; Liu *et al.*, 2008; Shen *et al.*, 1998) may lead to oncogenic activation. Since the molecular mechanism for AID targeting has not been elucidated, it is only speculative (but very likely) that any aberrant function in this system will induce cancer progression. It is also not known if the targeting or the DNA lesion resolution (see below) will be the determining factor in off-target activity. Interestingly, AIDs mis-targeting may be responsible for drug resistance in CML, by mutating the oncogenic translocation protein BCR-ABL (Klemm *et al.*, 2009).

8.2.4.6 *Substrate and activity*

Once at the target site, the type of substrate, substrate availability, enzymatic turnover and enzymatic kinetics can be a means of regulation and an alteration in AID's oncogenic potential. Enzymatic kinetics evolve with cellular need. Cytoplasmic DNA deaminases that inactivate foreign DNA should act efficiently on the substrate with less of a consequence for off-target activity. A nuclear DNA deaminase that can potentially mutate a number of important cellular genes should have more layers of control. The cytoplasmic DNA deaminase APOBEC3G stays associated with its substrate after its first deamination and moves in a processive 3′ to 5′ direction towards the next possible target (Chelico *et al.*, 2006; Coker and Petersen-Mahrt, 2007). Possibly to avoid oncogenic mutations, nuclear AID dissociates from its substrate between deaminations (but can re-associate with the same substrate; Coker and Petersen-Mahrt, 2007; Pham *et al.*, 2007), but the exact *in vivo* molecular mechanism at the target site still needs to be identified. Any alterations in AID kinetics can have substantial downstream effects.

Substrate availability plays an important role in DNA deamination. AID requires single-stranded DNA as a substrate, allowing for chromatin state and structure to provide a significant role in regulation, some of which can be achieved during transcription or replication. Clearly the 'natural state' of the target loci, e.g. R-loop structures in Ig switch regions (see Chapters 3 and 5), can predispose a locus for oncogenic events during AID activity (Yu *et al.*, 2005; Gomez-Gonzalez and Aguilera, 2007). These sites of fragile DNA have also been shown to be sites of oncogenic translocations (Tsai *et al.*, 2009), and may show a link between CpGs and translocations (Tsai *et al.*, 2008). Changes in epigenetic state are a hallmark of numerous cancers (Allis *et al.*, 2006), and germline mutations in key histone acetylases (CBP or p300) have been linked to tumor predisposition (Iyer *et al.*, 2004). Therefore, changes in DNA accessibility of tumor suppressors or oncogenes may directly lead to a change in AID substrate status.

Methylation of DNA has long been known to mark the genome for alterations in chromatin state and accessibility: if taking place near gene loci, multiple methylation of cytosines at carbon-5 (5meC) in the context of a CpG pair lead to gene inactivation (Allis *et al.*, 2006). On the other hand, de-methylation in the loci of silenced oncogenes (near transcription control regions) have been implicated in mis-activation of genes (Feinberg and Vogelstein, 1983; Ehrlich *et al.*, 1982). There are also examples of genome-wide hypomethylation in cancer cells, and enhanced mutations at CpGs in tumor samples (Lapeyre and Becker, 1979; Jones and Baylin, 2002). Mutations at CpGs (mostly transitions) are the most common form of point mutations in cancer, but at the same time our genomes only have one-fifth of the expected CpGs. Because AID can deaminate 5meC (Morgan *et al.*, 2004) – also noted but not commented on by Bransteitter and colleagues (Bransteitter *et al.*, 2003) – and is expressed during active epigenetic reprogramming (Morgan *et al.*, 2004), AID provides a direct link to the epigenome of a cell. Data from zebra fish indicated, that AID's ability to interact with enzymes from the base excision repair pathway can also change the epigenetic status during development (Rai *et al.*, 2008). Clearly, all of the aforementioned control pathways will also affect AID's ability to access and deaminate 5meC, and with it alter genomic expression of oncogenes and tumor suppressors.

8.2.4.7 *DNA lesion resolution*

Initially, it had been suggested that AID's activity on the Ig loci is the key step in achieving the high specificity of SHM. Subsequent analysis though, has indicated that the pathway in which the dU lesion is treated by DNA repair proteins contributes significantly to the outcome (see Chapters 5 and 6). Although AID seems to be targeted preferentially to the Ig locus, other loci are also mutated by AID. Importantly, once base excision repair (BER) and mismatch repair (MMR) had been removed from the system, the mutation frequency changed to different extents at different loci (Liu *et al.*, 2008; Liu and Schatz, 2009). This indicated that not all loci are treated equally in terms of repair fidelity, and could explain variable mutation frequency of oncogenes.

Furthermore, a co-upregulatory link in "professional" AID-expressing B cells has been observed between AID and various DNA repair proteins (Klemm *et al.*, 2009), but not in aberrantly AID-expressing cells of myeloid leukemia origin. These factors include those that were previously linked to suppression of translocations and other AID-induced genomic alterations, e.g. ATM (Ramiro *et al.*, 2006), BRCA1, FANCD2 and RAD51 (Longerich *et al.*, 2008), and could provide a protective role in B cells. This lack of a proper response machinery to AID-induced damage could have dramatic effects in "non-professional" AID-expressing cells, and would make AID an even more carcinogenic agent. In line with this, B cells have been shown to cope better with overexpressed AID, given that in a constitutive AID-expressing mouse model, surprisingly few B cell lymphomas arise in comparison to cancers of other origins (Okazaki *et al.*, 2003a).

Furthermore, it has been suggested that AID can interact with DNA repair proteins directly, such as the DNA damage response kinase DNA-PK (Wu *et al.*, 2005). Therefore, it is possible for AID to influence the DNA repair pathways downstream of its dU lesion, in effect "flagging" the lesion. Again, alterations in the DNA repair pathways or proteins will have oncogeneic consequences. It is almost self-evident that any alteration in a DNA repair protein has oncogenic potential, therefore, for details on how alterations in a DNA repair pathway or in a DNA repair protein can influence oncogenicity, the reader is asked to look at some of the recent reviews or excellent textbooks on this matter (Friedberg *et al.*, 2006; Jackson and Bartek, 2009).

8.3 Acknowledgements

This work was supported by grants from the Institut National de la Santé et de la Recherche Médicale, the Association de la Recherche contre le Cancer, the Association Nationale pour la Recherche (HIGM) and European Community (EURO-PADnet; to A.D.).

8.4 References

1. Al-Dhaheri M.H., Rowan B.G. (2007). Protein kinase A exhibits selective modulation of estradiol-dependent transcription in breast cancer cells that is associated with decreased ligand binding, altered estrogen receptor alpha promoter interaction, and changes in receptor phosphorylation. *Mol Endocrinol* **21**: 439–456.
2. Albesiano E., Messmer B.T., Damle R.N. *et al.* (2003). Activation-induced cytidine deaminase in chronic lymphocytic leukemia B cells: expression as multiple forms in a dynamic, variably sized fraction of the clone. *Blood* **102**: 3333–3339.
3. Allis C.D., Jenuwein T., Reinberg D. (2006). *Epigenetics*, Cold Spring Harbor, N.Y., Cold Spring Harbor Laboratory Press.

4. Aoufouchi S., Faili A., Zober C. *et al.* (2008). Proteasomal degradation restricts the nuclear lifespan of AID. *J Exp Med* **205**: 1357–1368.
5. Armitage R.J., Macduff B.M., Spriggs M.K. *et al.* (1993). Human B cell proliferation and Ig secretion induced by recombinant CD40 ligand are modulated by soluble cytokines. *J Immunol* **150**: 3671–3680.
6. Aruffo A., Farrington M., Hollenbaugh D. *et al.* (1993). The CD40 ligand, gp39, is defective in activated T cells from patients with X-linked hyper-IgM syndrome. *Cell* **72**: 291–300.
7. Babbage G., Garand R., Robillard N. *et al.* (2004). Mantle cell lymphoma with t(11;14) and unmutated or mutated VH genes expresses AID and undergoes isotype switch events. *Blood* **103**: 2795–2798.
8. Barreto V., Reina-San-Martin B., Ramiro A.R. *et al.* (2003). C-terminal deletion of AID uncouples class switch recombination from somatic hypermutation and gene conversion. *Mol Cell* **12**: 501–508.
9. Bartel D.P. (2004). MicroRNAs: genomics, biogenesis, mechanism, and function. *Cell* **116**: 281–297.
10. Benezra R., Rafii S., Lyden D. (2001). The Id proteins and angiogenesis. *Oncogene* **20**: 8334–8341.
11. Bijl J., van Oostveen J.W., Kreike M. *et al.* (1996). Expression of HOXC4, HOXC5, and HOXC6 in human lymphoid cell lines, leukemias, and benign and malignant lymphoid tissue. *Blood* **87**: 1737–1745.
12. Bijl J.J., Rieger E., van Oostveen J.W. *et al.* (1997). HOXC4, HOXC5, and HOXC6 expression in primary cutaneous lymphoid lesions. High expression of HOXC5 in anaplastic large-cell lymphomas. *Am J Pathol* **151**: 1067–1074.
13. Boikos S.A., Stratakis C.A. (2006). Carney complex: pathology and molecular genetics. *Neuroendocrinology* **83**: 189–199.
14. Bradbury A.W., Carter D.C., Miller W.R. *et al.* (1994). Protein kinase A (PK-A) regulatory subunit expression in colorectal cancer and related mucosa. *Br J Cancer* **69**: 738–742.
15. Bransteitter R., Pham P., Scharff M.D. *et al.* (2003). Activation-induced cytidine deaminase deaminates deoxycytidine on single-stranded DNA but requires the action of RNase. *Proc Natl Acad Sci USA* **100**: 4102–4107.
16. Brar S.S., Watson M., Diaz M. (2004). Activation-induced cytosine deaminase (AID) is actively exported out of the nucleus but retained by the induction of DNA breaks. *J Biol Chem* **279**: 26395–26401.
17. Bruns H.A., Kaplan M.H. (2006). The role of constitutively active Stat6 in leukemia and lymphoma. *Crit Rev Oncol Hematol* **57**: 245–253.
18. Busslinger M., Klix N., Pfeffer P. *et al.* (1996). Deregulation of PAX-5 by translocation of the Emu enhancer of the IgH locus adjacent to two alternative PAX-5 promoters in a diffuse large-cell lymphoma. *Proc Natl Acad Sci USA* **93**: 6129–6134.
19. Careccia S., Mainardi S., Pelosi A. *et al.* (2009). A restricted signature of miRNAs distinguishes APL blasts from normal promyelocytes. *Oncogene* **28**: 4034–4040.

20. Catalan N., Selz F., Imai K. *et al.* (2003). The block in immunoglobulin class switch recombination caused by activation-induced cytidine deaminase deficiency occurs prior to the generation of DNA double strand breaks in switch mu region. *J Immunol* **171**: 2504–2509.
21. Chelico L., Pham P., Calabrese P. *et al.* (2006). APOBEC3G DNA deaminase acts processively 3' → 5' on single-stranded DNA. *Nat Struct Mol Biol* **13**: 392–399.
22. Cobaleda C., Busslinger M. (2008). Developmental plasticity of lymphocytes. *Curr Opin Immunol* **20**: 139–148.
23. Coker H.A., Petersen-Mahrt S.K. (2007). The nuclear DNA deaminase AID functions distributively whereas cytoplasmic APOBEC3G has a processive mode of action. *DNA Repair* **6**: 235–243.
24. Conti A., Aguennouz M., La Torre D. *et al.* (2009). miR-21 and 221 upregulation and miR-181b downregulation in human grade II-IV astrocytic tumors. *J Neurooncol* **93**: 325–332.
25. Conticello S.G., Ganesh K., Xue K. *et al.* (2008). Interaction between antibody-diversification enzyme AID and spliceosome-associated factor CTNNBL1. *Mol Cell* **31**: 474–484.
26. Conticello S.G., Thomas C.J., Petersen-Mahrt S.K. *et al.* (2005). Evolution of the AID/APOBEC family of polynucleotide (deoxy)cytidine deaminases. *Mol Biol Evol* **22**: 367–377.
27. Cornford P., Evans J., Dodson A. *et al.* (1999). Protein kinase C isoenzyme patterns characteristically modulated in early prostate cancer. *Am J Pathol* **154**: 137–144.
28. Costinean S., Zanesi N., Pekarsky Y. *et al.* (2006). Pre-B cell proliferation and lymphoblastic leukemia/high-grade lymphoma in E(mu)-miR155 transgenic mice. *Proc Natl Acad Sci USA* **103**: 7024–7029.
29. Croce C.M. (2009). Causes and consequences of microRNA dysregulation in cancer. *Nat Rev Genet* **10**: 704–714.
30. de Bosscher K., van den Berghe W., Haegeman G. (2006). Cross-talk between nuclear receptors and nuclear factor kappaB. *Oncogene* **25**: 6868–6886.
31. de Yebenes V.G., Belver L., Pisano D.G. *et al.* (2008). miR-181b negatively regulates activation-induced cytidine deaminase in B cells. *J Exp Med* **205**: 2199–2206.
32. Dedeoglu F., Horwitz B., Chaudhuri J. *et al.* (2004). Induction of activation-induced cytidine deaminase gene expression by IL-4 and CD40 ligation is dependent on STAT6 and NFkappaB. *Int Immunol* **16**: 395–404.
33. Dijkman R., Tensen C.P., Buettner M. *et al.* (2006). Primary cutaneous follicle center lymphoma and primary cutaneous large B-cell lymphoma, leg type, are both targeted by aberrant somatic hypermutation but demonstrate differential expression of AID. *Blood* **107**: 4926–4929.
34. Ding L., Wang H., Lang W. *et al.* (2002). Protein kinase C-epsilon promotes survival of lung cancer cells by suppressing apoptosis through dysregulation of the mitochondrial caspase pathway. *J Biol Chem* **277**: 35305–35313.

35. Doi T., Kato L., Ito S. *et al.* (2009). The C-terminal region of activation-induced cytidine deaminase is responsible for a recombination function other than DNA cleavage in class switch recombination. *Proc Natl Acad Sci USA* **106**: 2758–2763.
36. Dorsett Y., McBride K.M., Jankovic M. *et al.* (2008). MicroRNA-155 suppresses activation-induced cytidine deaminase-mediated Myc-Igh translocation. *Immunity* **28**: 630–638.
37. Durandy A., Peron S., Taubenheim N. *et al.* (2006). Activation-induced cytidine deaminase: structure-function relationship as based on the study of mutants. *Hum Mutat* **27**: 1185–1191.
38. Durandy A., Schiff C., Bonnefoy J.Y. *et al.* (1993). Induction by anti-CD40 antibody or soluble CD40 ligand and cytokines of IgG, IgA and IgE production by B cells from patients with X-linked hyper IgM syndrome. *Eur J Immunol* **23**: 2294–2299.
39. Ehrlich M., Gama-Sosa M.A., Huang L.H. *et al.* (1982). Amount and distribution of 5-methylcytosine in human DNA from different types of tissues of cells. *Nucleic Acids Res* **10**: 2709–2721.
40. Eis P.S., Tam W., Sun L. *et al.* (2005). Accumulation of miR-155 and BIC RNA in human B cell lymphomas. *Proc Natl Acad Sci USA* **102**: 3627–3632.
41. Endo Y., Marusawa H., Kinoshita K. *et al.* (2007). Expression of activation-induced cytidine deaminase in human hepatocytes via NF-kappaB signaling. *Oncogene* **26**: 5587–5595.
42. Engels K., Jungnickel B., Tobollik S. *et al.* (2008). Expression of activation-induced cytidine deaminase in malignant lymphomas infiltrating the bone marrow. *Appl Immunohistochem Mol Morphol* **16**: 521–529.
43. Epeldegui M., Hung Y.P., McQuay A. *et al.* (2007). Infection of human B cells with Epstein-Barr virus results in the expression of somatic hypermutation-inducing molecules and in the accrual of oncogene mutations. *Mol Immunol* **44**: 934–942.
44. Feinberg A.P., Vogelstein B. (1983). Hypomethylation distinguishes genes of some human cancers from their normal counterparts. *Nature* **301**: 89–92.
45. Feldhahn N., Henke N., Melchior K. *et al.* (2007). Activation-induced cytidine deaminase acts as a mutator in BCR-ABL1-transformed acute lymphoblastic leukemia cells. *J Exp Med* **204**: 1157–1166.
46. Ferrari S., Giliani S., Insalaco A. *et al.* (2001). Mutations of CD40 gene cause an autosomal recessive form of immunodeficiency with hyper IgM. *Proc Natl Acad Sci USA* **98**: 12614–12619.
47. Forconi F., Sahota S.S., Raspadori D. *et al.* (2004). Hairy cell leukemia: at the crossroad of somatic mutation and isotype switch. *Blood* **104**: 3312–3317.
48. Friedberg E.C., Aguilera A., Gellert M. *et al.* (2006). DNA repair: from molecular mechanism to human disease. *DNA Repair* **5**: 986–996.
49. Geisberger R., Rada C., Neuberger M.S. (2009). The stability of AID and its function in class-switching are critically sensitive to the identity of its nuclear-export sequence. *Proc Natl Acad Sci USA* **106**: 6736–6741.

50. Gil Y., Levy-Nabot S., Steinitz M. *et al.* (2007). Somatic mutations and activation-induced cytidine deaminase (AID) expression in established rheumatoid factor-producing lymphoblastoid cell line. *Mol Immunol* **44**: 494–505.
51. Gluz O., Wild P., Meiler R. *et al.* (2008). Nuclear karyopherin alpha2 expression predicts poor survival in patients with advanced breast cancer irrespective of treatment intensity. *Int J Cancer* **123**: 1433–1438.
52. Gomez-Gonzalez B., Aguilera A. (2007). Activation-induced cytidine deaminase action is strongly stimulated by mutations of the THO complex. *Proc Natl Acad Sci USA* **104**: 8409–8414.
53. Gonda H., Sugai M., Nambu Y. *et al.* (2003). The balance between Pax5 and Id2 activities is the key to AID gene expression. *J Exp Med* **198**: 1427–1437.
54. Gourzi P., Leonova T., Papavasiliou F.N. (2006). A role for activation-induced cytidine deaminase in the host response against a transforming retrovirus. *Immunity* **24**: 779–786.
55. Greenman C., Stephens P., Smith R. *et al.* (2007). Patterns of somatic mutation in human cancer genomes. *Nature* **446**: 153–158.
56. Greeve J., Philipsen A., Krause K. *et al.* (2003). Expression of activation-induced cytidine deaminase in human B-cell non-Hodgkin lymphomas. *Blood* **101**: 3574–3580.
57. Greiner A., Tobollik S., Buettner M. *et al.* (2005). Differential expression of activation-induced cytidine deaminase (AID) in nodular lymphocyte-predominant and classical Hodgkin lymphoma. *J Pathol* **205**: 541–547.
58. Guikema J.E., Rosati S., Akkermans K. *et al.* (2005). Quantitative RT-PCR analysis of activation-induced cytidine deaminase expression in tissue samples from mantle cell lymphoma and B-cell chronic lymphocytic leukemia patients. *Blood* **105**: 2997–2998.
59. Hamajima N., Naito M., Kondo T. *et al.* (2006). Genetic factors involved in the development of Helicobacter pylori-related gastric cancer. *Cancer Sci* **97**: 1129–1138.
60. Hardianti M.S., Tatsumi E., Syampurnawati M. *et al.* (2004). Activation-induced cytidine deaminase expression in follicular lymphoma: association between AID expression and ongoing mutation in FL. *Leukemia* **18**: 826–831.
61. Hardianti M.S., Tatsumi E., Syampurnawati M. *et al.* (2005). Presence of somatic hypermutation and activation-induced cytidine deaminase in acute lymphoblastic leukemia L2 with t(14;18)(q32;q21). *Eur J Haematol* **74**: 11–19.
62. Haughian J.M., Jackson T.A., Koterwas D.M. *et al.* (2006). Endometrial cancer cell survival and apoptosis is regulated by protein kinase C alpha and delta. *Endocr Relat Cancer* **13**: 1251–1267.
63. Hoeller D., Dikic I. (2009). Targeting the ubiquitin system in cancer therapy. *Nature* **458**: 438–444.
64. Hoeller D., Hecker C.M., Dikic I. (2006). Ubiquitin and ubiquitin-like proteins in cancer pathogenesis. *Nat Rev Cancer* **6**: 776–788.
65. Iavarone A., Lasorella A. (2004). Id proteins in neural cancer. *Cancer Lett* **204**: 189–196.

66. Iida S., Rao P.H., Nallasivam P. *et al.* (1996). The t(9;14)(p13;q32) chromosomal translocation associated with lymphoplasmacytoid lymphoma involves the PAX-5 gene. *Blood* **88**: 4110–4117.
67. Imai K., Zhu Y., Revy P. *et al.* (2005). Analysis of class switch recombination and somatic hypermutation in patients affected with autosomal dominant hyper-IgM syndrome type 2. *Clin Immunol* **115**: 277–285.
68. Ito S., Nagaoka H., Shinkura R. *et al.* (2004). Activation-induced cytidine deaminase shuttles between nucleus and cytoplasm like apolipoprotein B mRNA editing catalytic polypeptide 1. *Proc Natl Acad Sci USA* **101**: 1975–1980.
69. Iyer N.G., Ozdag H., Caldas C. (2004). p300/CBP and cancer. *Oncogene* **23**: 4225–4231.
70. Jackson S.P., Bartek J. (2009). The DNA-damage response in human biology and disease. *Nature* **461**: 1071–1078.
71. Jones P.A., Baylin S.B. (2002). The fundamental role of epigenetic events in cancer. *Nat Rev Genet* **3**: 415–428.
72. Katagiri Y., Hozumi Y., Kondo S. (2006). Knockdown of Skp2 by siRNA inhibits melanoma cell growth in vitro and in vivo. *J Dermatol Sci* **42**: 215–224.
73. Klein U., Rajewsky K., Kuppers R. (1998). Human immunoglobulin (Ig)M+IgD+ peripheral blood B cells expressing the CD27 cell surface antigen carry somatically mutated variable region genes: CD27 as a general marker for somatically mutated (memory) B cells. *J Exp Med* **188**: 1679–1689.
74. Klemm L., Duy C., Iacobucci I. *et al.* (2009). The B cell mutator AID promotes B lymphoid blast crisis and drug resistance in chronic myeloid leukemia. *Cancer Cell* **16**: 232–245.
75. Kluiver J., Haralambieva E., De Jong D. *et al.* (2006). Lack of BIC and microRNA miR-155 expression in primary cases of Burkitt lymphoma. *Genes Chromosomes Cancer* **45**: 147–153.
76. Kluiver J., Poppema S., de Jong D. *et al.* (2005). BIC and miR-155 are highly expressed in Hodgkin, primary mediastinal and diffuse large B cell lymphomas. *J Pathol* **207**: 243–249.
77. Kluiver J., van Den Berg A., De Jong D. *et al.* (2007). Regulation of pri-microRNA BIC transcription and processing in Burkitt lymphoma. *Oncogene* **26**: 3769–3776.
78. Korthauer U., Graf D., Mages H.W. *et al.* (1993). Defective expression of T-cell CD40 ligand causes X-linked immunodeficiency with hyper-IgM. *Nature* **361**: 539–541.
79. Kou T., Marusawa H., Kinoshita K. *et al.* (2007). Expression of activation-induced cytidine deaminase in human hepatocytes during hepatocarcinogenesis. *Int J Cancer* **120**: 469–476.
80. Kuppers R. (2005). Mechanisms of B-cell lymphoma pathogenesis. *Nat Rev Cancer* **5**: 251–262.
81. Kusunoki M., Sakanoue Y., Hatada T. *et al.* (1992). Protein kinase C activity in human colonic adenoma and colorectal carcinoma. *Cancer* **69**: 24–30.
82. Lahn M., Kohler G., Sundell K. *et al.* (2004). Protein kinase C alpha expression in breast and ovarian cancer. *Oncology* **67**: 1–10.

83. Lapeyre J.N., Becker F.F. (1979). 5-Methylcytosine content of nuclear DNA during chemical hepatocarcinogenesis and in carcinomas which result. *Biochem Biophys Res Commun* **87**: 698–705.
84. Lasorella A., Uo T., Iavarone A. (2001). Id proteins at the cross-road of development and cancer. *Oncogene* **20**: 8326–8333.
85. Lee W.I., Torgerson T.R., Schumacher M.J. *et al.* (2005). Molecular analysis of a large cohort of patients with the hyper immunoglobulin M (IgM) syndrome. *Blood* **105**: 1881–1890.
86. Leung A.K., Sharp P.A. (2006). Function and localization of microRNAs in mammalian cells. *Cold Spring Harb Symp Quant Biol* **71**: 29–38.
87. Lim M.S., Adamson A., Lin Z. *et al.* (2002). Expression of Skp2, a p27(Kip1) ubiquitin ligase, in malignant lymphoma: correlation with p27(Kip1) and proliferation index. *Blood* **100**: 2950–2956.
88. Lindahl T. (1993). Instability and decay of the primary structure of DNA. *Nature* **362**: 709–715.
89. Liu M., Duke J.L., Richter D.J. *et al.* (2008). Two levels of protection for the B cell genome during somatic hypermutation. *Nature* **451**: 841–845.
90. Liu M., Schatz D.G. (2009). Balancing AID and DNA repair during somatic hypermutation. *Trends Immunol* **30**: 173–181.
91. Longerich S., Orelli B.J., Martin R.W. *et al.* (2008). Brca1 in immunoglobulin gene conversion and somatic hypermutation. *DNA Repair* **7**: 253–266.
92. Lossos I.S., Levy R., Alizadeh A.A. (2004). AID is expressed in germinal center B-cell-like and activated B-cell-like diffuse large-cell lymphomas and is not correlated with intraclonal heterogeneity. *Leukemia* **18**: 1775–1779.
93. Machida K., Cheng K.T., Sung V.M. *et al.* (2004). Hepatitis C virus induces a mutator phenotype: enhanced mutations of immunoglobulin and protooncogenes. *Proc Natl Acad Sci USA* **101**: 4262–4267.
94. Maeda S., Yoshida H., Mitsuno Y. *et al.* (2002). Analysis of apoptotic and antiapoptotic signalling pathways induced by Helicobacter pylori. *Mol Pathol* **55**: 286–293.
95. Matsumoto Y., Marusawa H., Kinoshita K. *et al.* (2007). Helicobacter pylori infection triggers aberrant expression of activation-induced cytidine deaminase in gastric epithelium. *Nat Med* **13**: 470–476.
96. Maul R.W., Gearhart P.J. (2009). Women, autoimmunity, and cancer: a dangerous liaison between estrogen and activation-induced deaminase? *J Exp Med* **206**: 11–13.
97. McBride K.M., Barreto V., Ramiro A.R. *et al.* (2004). Somatic hypermutation is limited by CRM1-dependent nuclear export of activation-induced deaminase. *J Exp Med* **199**: 1235–1244.
98. McCann K.J., Ashton-Key M., Smith K. *et al.* (2009). Primary central nervous system lymphoma: tumor-related clones exist in the blood and bone marrow with evidence for separate development. *Blood* **113**: 4677–4680.
99. Miller G.J., Miller H.L., Van Bokhoven A. *et al.* (2003). Aberrant HOXC expression accompanies the malignant phenotype in human prostate. *Cancer Res* **63**: 5879–5888.

100. Minegishi Y., Lavoie A., Cunningham-Rundles C. *et al.* (2000). Mutations in activation-induced cytidine deaminase in patients with hyper IgM syndrome. *Clin Immunol* **97**: 203–210.
101. Mischak H., Goodnight J.A., Kolch W. *et al.* (1993). Overexpression of protein kinase C-delta and -epsilon in NIH 3T3 cells induces opposite effects on growth, morphology, anchorage dependence, and tumorigenicity. *J Biol Chem* **268**: 6090–6096.
102. Moldenhauer G., Popov S.W., Wotschke B. *et al.* (2006). AID expression identifies interfollicular large B cells as putative precursors of mature B-cell malignancies. *Blood* **107**: 2470–2473.
103. Montesinos-Rongen M., Schmitz R., Courts C. *et al.* (2005). Absence of immunoglobulin class switch in primary lymphomas of the central nervous system. *Am J Pathol* **166**: 1773–1779.
104. Morgan H.D., Dean W., Coker H.A. *et al.* (2004). Activation-induced cytidine deaminase deaminates 5-methylcytosine in DNA and is expressed in pluripotent tissues: implications for epigenetic reprogramming. *J Biol Chem* **279**: 52353–52360.
105. Morrison A.M., Jager U., Chott A. *et al.* (1998). Deregulated PAX-5 transcription from a translocated IgH promoter in marginal zone lymphoma. *Blood* **92**: 3865–3878.
106. Murray N.R., Davidson L.A., Chapkin R.S. *et al.* (1999). Overexpression of protein kinase C betaII induces colonic hyperproliferation and increased sensitivity to colon carcinogenesis. *J Cell Biol* **145**: 699–711.
107. Muschen M., Re D., Jungnickel B. *et al.* (2000). Somatic mutation of the CD95 gene in human B cells as a side-effect of the germinal center reaction. *J Exp Med* **192**: 1833–1840.
108. Mutka S.C., Yang W.Q., Dong S.D. *et al.* (2009). Identification of nuclear export inhibitors with potent anticancer activity in vivo. *Cancer Res* **69**: 510–517.
109. Nakajima G., Hayashi K., Xi Y. *et al.* (2006). Non-coding microRNAs hsa-let-7g and hsa-miR-181b are associated with chemoresponse to S-1 in colon cancer. *Cancer Genomics Proteomics* **3**: 317–324.
110. Navarro A., Bea S., Fernandez V. *et al.* (2009). MicroRNA expression, chromosomal alterations, and immunoglobulin variable heavy chain hypermutations in Mantle cell lymphomas. *Cancer Res* **69**: 7071–7078.
111. Negrini M., Ferracin M., Sabbioni S. *et al.* (2007). MicroRNAs in human cancer: from research to therapy. *J Cell Sci* **120**: 1833–1840.
112. Ni Z., Lou W., Lee S.O. *et al.* (2002). Selective activation of members of the signal transducers and activators of transcription family in prostate carcinoma. *J Urol* **167**: 1859–1862.
113. O'Brian C., Vogel V.G., Singletary S.E. *et al.* (1989). Elevated protein kinase C expression in human breast tumor biopsies relative to normal breast tissue. *Cancer Res* **49**: 3215–3217.
114. Okazaki I., Yoshikawa K., Kinoshita K. *et al.* (2003a). Activation-induced cytidine deaminase links class switch recombination and somatic hypermutation. *Ann NY Acad Sci* **987**: 1–8.

115. Okazaki I.M., Hiai H., Kakazu N. *et al.* (2003b). Constitutive expression of AID leads to tumorigenesis. *J Exp Med* **197**: 1173–1181.
116. Pallasch C.P., Patz M., Park Y.J. *et al.* (2009). miRNA deregulation by epigenetic silencing disrupts suppression of the oncogene PLAG1 in chronic lymphocytic leukemia. *Blood* **114**: 3255–3264.
117. Park S.R., Zan H., Pal Z. *et al.* (2009). HoxC4 binds to the promoter of the cytidine deaminase AID gene to induce AID expression, class-switch DNA recombination and somatic hypermutation. *Nat Immunol* **10**: 540–550.
118. Pasqualucci L., Guglielmino R., Houldsworth J. *et al.* (2004). Expression of the AID protein in normal and neoplastic B cells. *Blood* **104**: 3318–3325.
119. Pasqualucci L., Migliazza A., Fracchiolla N. *et al.* (1998). BCL-6 mutations in normal germinal center B cells: evidence of somatic hypermutation acting outside Ig loci. *Proc Natl Acad Sci USA* **95**: 11816–11821.
120. Pasqualucci L., Neumeister P., Goossens T. *et al.* (2001). Hypermutation of multiple proto-oncogenes in B-cell diffuse large-cell lymphomas. *Nature* **412**: 341–346.
121. Patenaude A.M., Orthwein A., Hu Y. *et al.* (2009). Active nuclear import and cytoplasmic retention of activation-induced deaminase. *Nat Struct Mol Biol* **16**: 517–527.
122. Pauklin S., Sernandez I.V., Bachmann G. *et al.* (2009). Estrogen directly activates AID transcription and function. *J Exp Med* **206**: 99–111.
123. Peron S., Pan-Hammarstrom Q., Imai K. *et al.* (2007). A primary immunodeficiency characterized by defective immunoglobulin class switch recombination and impaired DNA repair. *J Exp Med* **204**: 1207–1216.
124. Petersen-Mahrt S.K., Coker H.A., Pauklin S. (2009). DNA deaminases: AIDing hormones in immunity and cancer. *J Mol Med* **87**: 893–897.
125. Pham P., Chelico L., Goodman M.F. (2007). DNA deaminases AID and APOBEC3G act processively on single-stranded DNA. *DNA Repair* **6**: 689–692.
126. Platet N., Prevostel C., Derocq D. *et al.* (1998). Breast cancer cell invasiveness: correlation with protein kinase C activity and differential regulation by phorbol ester in estrogen receptor-positive and -negative cells. *Int J Cancer* **75**: 750–756.
127. Pollack A., Bae K., Khor L.Y. *et al.* (2009). The importance of protein kinase A in prostate cancer: relationship to patient outcome in Radiation Therapy Oncology Group trial 92-02. *Clin Cancer Res* **15**: 5478–5484.
128. Popov S.W., Moldenhauer G., Wotschke B. *et al.* (2007). Target sequence accessibility limits activation-induced cytidine deaminase activity in primary mediastinal B-cell lymphoma. *Cancer Res* **67**: 6555–6564.
129. Poppe B., De Paepe P., Michaux L. *et al.* (2005). PAX5/IGH rearrangement is a recurrent finding in a subset of aggressive B-NHL with complex chromosomal rearrangements. *Genes Chromosomes Cancer* **44**: 218–223.
130. Quartier P., Bustamante J., Sanal O. *et al.* (2004). Clinical, immunologic and genetic analysis of 29 patients with autosomal recessive hyper-IgM syndrome due to activation-induced cytidine deaminase deficiency. *Clin Immunol* **110**: 22–29.
131. Rabbitts T.H., Forster A., Hamlyn P. *et al.* (1984). Effect of somatic mutation within translocated c-myc genes in Burkitt's lymphoma. *Nature* **309**: 592–597.

132. Rada C., Jarvis J.M., Milstein C. (2002). AID-GFP chimeric protein increases hypermutation of Ig genes with no evidence of nuclear localization. *Proc Natl Acad Sci USA* **99**: 7003–7008.
133. Rai K., Huggins I.J., James S.R. *et al.* (2008). DNA demethylation in zebrafish involves the coupling of a deaminase, a glycosylase, and gadd45. *Cell* **135**: 1201–1212.
134. Ramiro A.R., Jankovic M., Callen E. *et al.* (2006). Role of genomic instability and p53 in AID-induced c-myc-Igh translocations. *Nature* **440**: 105–109.
135. Ramiro A.R., Jankovic M., Eisenreich T. *et al.* (2004). AID is required for c-myc/IgH chromosome translocations in vivo. *Cell* **118**: 431–438.
136. Reiniger L., Bodor C., Bognar A. *et al.* (2006). Richter's and prolymphocytic transformation of chronic lymphocytic leukemia are associated with high mRNA expression of activation-induced cytidine deaminase and aberrant somatic hypermutation. *Leukemia* **20**: 1089–1095.
137. Revy P., Muto T., Levy Y. *et al.* (2000). Activation-induced cytidine deaminase (AID) deficiency causes the autosomal recessive form of the Hyper-IgM syndrome (HIGM2). *Cell* **102**: 565–575.
138. Ritz O., Guiter C., Castellano F. *et al.* (2009). Recurrent mutations of the STAT6 DNA binding domain in primary mediastinal B-cell lymphoma. *Blood* **114**: 1236–1242.
139. Robbiani D.F., Bothmer A., Callen E. *et al.* (2008). AID is required for the chromosomal breaks in c-myc that lead to c-myc/IgH translocations. *Cell* **135**: 1028–1038.
140. Robbiani D.F., Bunting S., Feldhahn N. *et al.* (2009). AID produces DNA double-strand breaks in non-Ig genes and mature B cell lymphomas with reciprocal chromosome translocations. *Mol Cell* **36**: 631–641.
141. Rosenberg B.R., Papavasiliou F.N. (2007). Beyond SHM and CSR: AID and related cytidine deaminases in the host response to viral infection. *Adv Immunol* **94**: 215–244.
142. Ruminy P., Jardin F., Picquenot J.M. *et al.* (2008). S(mu) mutation patterns suggest different progression pathways in follicular lymphoma: early direct or late from FL progenitor cells. *Blood* **112**: 1951–1959.
143. Sachidanandam R., Weissman D., Schmidt S.C. *et al.* (2001). A map of human genome sequence variation containing 1.42 million single nucleotide polymorphisms. *Nature* **409**: 928–933.
144. Sayegh C.E., Quong M.W., Agata Y. *et al.* (2003). E-proteins directly regulate expression of activation-induced deaminase in mature B cells. *Nat Immunol* **4**: 586–593.
145. Schmitz K.J., Hey S., Schinwald A. *et al.* (2009). Differential expression of microRNA 181b and microRNA 21 in hyperplastic polyps and sessile serrated adenomas of the colon. *Virchows Arch* **455**: 49–54.
146. Shen H.M., Peters A., Baron B. *et al.* (1998). Mutation of BCL-6 gene in normal B cells by the process of somatic hypermutation of Ig genes. *Science* **280**: 1750–1752.

147. Shi L., Cheng Z., Zhang J. *et al.* (2008). hsa-mir-181a and hsa-mir-181b function as tumor suppressors in human glioma cells. *Brain Res* **1236**: 185–193.
148. Shinkura R., Tian M., Smith M. *et al.* (2003). The influence of transcriptional orientation on endogenous switch region function. *Nat Immunol* **4**: 435–441.
149. Skinnider B.F., Elia A.J., Gascoyne R.D. *et al.* (2002). Signal transducer and activator of transcription 6 is frequently activated in Hodgkin and Reed-Sternberg cells of Hodgkin lymphoma. *Blood* **99**: 618–626.
150. Smit L.A., Bende R.J., Aten J. *et al.* (2003). Expression of activation-induced cytidine deaminase is confined to B-cell non-Hodgkin's lymphomas of germinal-center phenotype. *Cancer Res* **63**: 3894–3898.
151. Solomon S.S., Majumdar G., Martinez-Hernandez A. *et al.* (2008). A critical role of Sp1 transcription factor in regulating gene expression in response to insulin and other hormones. *Life Sci* **83**: 305–312.
152. Srebrow A., Kornblihtt A.R. (2006). The connection between splicing and cancer. *J Cell Sci* **119**: 2635–2641.
153. Stice J.P., Knowlton A.A. (2008). Estrogen, NFkappaB, and the heat shock response. *Mol Med* **14**: 517–527.
154. Ta V.T., Nagaoka H., Catalan N. *et al.* (2003). AID mutant analyses indicate requirement for class-switch-specific cofactors. *Nat Immunol* **4**: 843–848.
155. Tachado S.D., Mayhew M.W., Wescott G.G. *et al.* (2002). Regulation of tumor invasion and metastasis in protein kinase C epsilon-transformed NIH3T3 fibroblasts. *J Cell Biochem* **85**: 785–797.
156. Takizawa M., Tolarova H., Li Z. *et al.* (2008). AID expression levels determine the extent of cMyc oncogenic translocations and the incidence of B cell tumor development. *J Exp Med* **205**: 1949–1957.
157. Tam W., Hughes S.H., Hayward W.S. *et al.* (2002). Avian bic, a gene isolated from a common retroviral site in avian leukosis virus-induced lymphomas that encodes a noncoding RNA, cooperates with c-myc in lymphomagenesis and erythroleukemogenesis. *J Virol* **76**: 4275–4286.
158. Teng G., Hakimpour P., Landgraf P. *et al.* (2008). MicroRNA-155 is a negative regulator of activation-induced cytidine deaminase. *Immunity* **28**: 621–629.
159. Tsai A.G., Engelhart A.E., Hatmal M.M. *et al.* (2009). Conformational variants of duplex DNA correlated with cytosine-rich chromosomal fragile sites. *J Biol Chem* **284**: 7157–7164.
160. Tsai A.G., Lu H., Raghavan S.C. *et al.* (2008). Human chromosomal translocations at CpG sites and a theoretical basis for their lineage and stage specificity. *Cell* **135**: 1130–1142.
161. Turner J.G., Sullivan D.M. (2008). CRM1-mediated nuclear export of proteins and drug resistance in cancer. *Curr Med Chem* **15**: 2648–2655.
162. van Maldegem F., Scheeren F.A., Aarti Jibodh R. *et al.* (2009). AID splice variants lack deaminase activity. *Blood* **113**: 1862–1864.
163. Vigorito E., Perks K.L., Abreu-Goodger C. *et al.* (2007). microRNA-155 regulates the generation of immunoglobulin class-switched plasma cells. *Immunity* **27**: 847–859.

164. Vuong B.Q., Lee M., Kabir S. *et al.* (2009). Specific recruitment of protein kinase A to the immunoglobulin locus regulates class-switch recombination. *Nat Immunol* **10**: 420–426.
165. Weinberg R.A. (2007). *The Biology of Cancer*, New York, London, Garland Science.
166. Willenbrock K., Renne C., Rottenkolber M. *et al.* (2009). The expression of activation induced cytidine deaminase in follicular lymphoma is independent of prognosis and stage. *Histopathology* **54**: 509–512.
167. Wu X., Darce J.R., Chang S.K. *et al.* (2008). Alternative splicing regulates activation-induced cytidine deaminase (AID): implications for suppression of AID mutagenic activity in normal and malignant B cells. *Blood* **112**: 4675–4682.
168. Wu X., Geraldes P., Platt J.L. *et al.* (2005). The double-edged sword of activation-induced cytidine deaminase. *J Immunol* **174**: 934–941.
169. Yadav A., Olaru A., Saltis M. *et al.* (2006). Identification of a ubiquitously active promoter of the murine activation-induced cytidine deaminase (AICDA) gene. *Mol Immunol* **43**: 529–541.
170. Yamagata K., Fujiyama S., Ito S. *et al.* (2009). Maturation of microRNA is hormonally regulated by a nuclear receptor. *Mol Cell* **36**: 340–347.
171. Yu K., Roy D., Bayramyan M. *et al.* (2005). Fine-structure analysis of activation-induced deaminase accessibility to class switch region R-loops. *Mol Cell Biol* **25**: 1730–1736.
172. Zhang J., Anastasiadis P.Z., Liu Y. *et al.* (2004). Protein kinase C (PKC) betaII induces cell invasion through a Ras/Mek-, PKC iota/Rac 1-dependent signaling pathway. *J Biol Chem* **279**: 22118–22123.
173. Zhu W., Qin W., Atasoy U. *et al.* (2009). Circulating microRNAs in breast cancer and healthy subjects. *BMC Res Notes* **2**: 89.

Chapter 9

AID in Aging and in Autoimmune Disease

Marilyn Diaz,[1] *Daniela Frasca,*[2] *and Bonnie Blomberg*[2]

[1]*Laboratory of Molecular Genetics, D3-01*
National Institute of Environmental Health Sciences
Research Triangle Park, NC, 27709
E-mail: diaz@niehs.nih.gov

[2]*Department of Microbiology and Immunology*
University of Miami School of Medicine
Miami, FL 33136
E-mail: bblomber@med.miami.edu

A surprising finding in recent years was the discovery that even moderate changes in AID levels, either through alleles carrying function-altering mutations in heterozygotes or through changes in its regulation, have a significant impact in class switch recombination and somatic hypermutation. In this chapter, we examine the impact of the alteration of AID levels in two aspects of human health: in aging and its potential association to age-related immunodeficiency, and in autoimmune disease. In the first section, evidence is discussed that strongly implicates decreased AID levels as a contributing factor for the suboptimal humoral responses in aged individuals. The second part of this chapter discusses the impact of altering AID levels in autoimmunity. Interestingly, any alteration in the levels of this protein in humans, in either direction, appears to be associated with a higher incidence of autoimmunity. This is not the case in mice, where lower levels of AID are associated with a decrease in autoimmunity. This discordance between laboratory mice and humans may be related to a high microbial burden in immunodeficient humans.

9.1 AID and Aging

Humoral and cellular immune responses are decreased by age in both humans and mice (Hodes, 1997; LeMaoult *et al.*, 1997; Pawelec *et al.*, 2002; Linton and Dorshkind, 2004; Sadighi Akha and Miller, 2005), leading to increased frequency and severity of infectious diseases and reduced protective effects of vaccination. Not only is the production of high-affinity protective antibodies decreased by age, but also the duration of protective immunity following immunization has been described to be shortened (Steger *et al.*, 1996; Weksler and Szabo, 2000). The decreased ability of aged individuals to produce high-affinity protective antibodies against infectious agents is likely to result from combined defects in T cells, B cells and antigen-presenting cells (APCs). Studies conducted in mice have shown that T cell help is diminished by age (Klinman and Kline, 1997; Linton and Thoman, 2001; Albright and Albright, 2003) and T cell-mediated suppression is increased (Sharma *et al.*, 2006). Memory T cell responses generated by naive T cells from young mice are functioning well even one year after priming; conversely, memory T cell responses generated in old age are defective both *in vitro* and *in vivo*, suggesting that naive CD4 T cells from aged mice are defective in generating good memory (Haynes *et al.*, 2003). This has been argued to explain the reduced antigen-specific B cell expansion, germinal center (GC) development and IgG production observed in aging (Haynes *et al.*, 2003). Work from our laboratory (see below) has identified intrinsic B cell defects with age in both mice and humans. Regulation of B cell function in GC occurs via costimulation (through CD40, CD80, CD86) and cytokine production (e.g. IL-4) from CD4 T cells. In both mice and humans the percentage and level of CD28 on CD4 T cells is not decreased (Peres *et al.*, 2003; Connoy *et al.*, 2006) but stimulatory signal pathways of CD4 T cells, and hence their cytokine production, are decreased with age (Engwerda *et al.*, 1994; Garcia and Miller, 2002).

We (Frasca *et al.*, 2003; Frasca *et al.*, 2004a; Frasca *et al.*, 2004b; Frasca *et al.*, 2008a) and others (Cancro *et al.*, 2009; Gibson *et al.*, 2009) have shown that intrinsic B cell defects occur in aging. These include decreases in Ig class switch recombination (CSR), activation-induced cytidine deaminase (AID) and E47 transcription factor (Frasca *et al.*,

2004b). Effects on somatic hypermutation (SHM) as well as on antibody affinity maturation have been varied depending on the system studied (Miller and Kelsoe, 1995; Yang *et al.*, 1996; Rogerson *et al.*, 2003). Moreover, decreased numbers of plasma cells producing both low- and high-affinity antibodies as a consequence of a recent antigen stimulation have also been reported (Han *et al.*, 2003).

9.2 Aging Decreases Humoral Immune Responses

In humans, aging affects antibody production both quantitatively and qualitatively with reduced serum concentrations of antigen-specific immunoglobulin (Ig), antibody specificity, affinity and CSR (Frasca *et al.*, 2008a; Frasca *et al.*, 2008b). There are decreased numbers of total B cells (Chong *et al.*, 2005; Shi *et al.*, 2005; Frasca *et al.*, 2008a) as well as, specifically, switch memory B cells in aged individuals (Frasca *et al.*, 2008a). Antibodies made by B lymphocytes prevent colonization by pathogens and provide immunity against invading pathogens. In the elderly and in patients with weakened immune systems, vaccines and therapeutic agents are likely to be not as effective because the number of functional B lymphocytes are fewer and Ig levels are reduced compared to healthy people. The changes in the humoral immune response during aging significantly contribute to the increased susceptibility of the elderly to infectious diseases and reduced response to vaccination (Gardner *et al.*, 2006; Frasca *et al.*, 2008b; Sambhara and McElhaney, 2009).

The decrease in CSR seen with age in mice (Frasca *et al.*, 2004b) and humans (Frasca *et al.*, 2008a) results in less IgG for an optimal newly-generated antigen response; not only are the IgG constant region effector functions critical for an optimal response but the switch and maturation to IgG is also associated with SHM and affinity maturation of the variable (V) region, both dependent on AID. CSR requires chromatin opening of a particular switch (S) region, and is mediated by cytokine-induced germline transcription. In this process, AID is required (Okazaki *et al.*, 2002; Nussenzweig and Alt, 2004). AID initiates CSR by deaminating cytidine residues in S regions, thus creating uracils, and the resulting mismatches are recognized by specific enzymes and excised, leading to DNA double-strand breaks (Rada *et al.*, 2002; Nussenzweig and Alt, 2004); see references therein, this book and Honjo and colleagues for an alternative model (Muramatsu *et al.*, 2007). In humans, mutations in the *Aicda* gene (encoding the AID protein) are associated with the absence of secondary antibodies and SHM, and produce hyper-IgM syndrome (Durandy, 2002), a disease associated with increased susceptibility to infections (Notarangelo *et al.*, 2006). AID can also induce point mutations in oncogenes such as *Bcl-6* (Pasqualucci *et al.*, 1998) and DNA double-strand breaks in Ig that are recognized as substrates for chromosome translocations (Dorsett *et al.*, 2007). DNA damage by AID is minimized in part because AID expression is restricted to GC B cells expressing Pax-5 (Gonda *et al.*, 2003) and E47 (Sayegh *et al.*, 2003) transcription factors and AID levels are regulated post-transcriptionally. This post-transcriptional regulation includes microRNA-155, which controls the half-life of AID mRNA (Dorsett *et al.*, 2008; Teng *et al.*, 2008) and also phosphorylation of the AID protein (Basu *et al.*, 2005; Pasqualucci *et al.*, 2006). Moreover, the concentration of AID in the nucleus is influenced by the amount of DNA breaks (Brar *et al.*, 2004). The *Aicda* gene is transcriptionally-regulated at least substantially by E proteins (Sayegh *et al.*, 2003; Tran *et al.*, 2010), which are class I basic helix loop helix (bHLH) proteins, first identified based on their ability to bind with relatively high affinity to the palindromic DNA sequence CANNTG, referred to as an E-box site (Ephrussi *et al.*, 1985; Henthorn *et al.*, 1990; Quong *et al.*, 2002). E-boxes have been found in the promoter and enhancer regions of many B lineage-specific genes and regulate a large number of processes involved in B cell commitment and differentiation (Murre *et al.*, 1989; Schlissel *et al.*, 1991; Sigvardsson *et al.*, 1997; Massari and Murre, 2000; Kee *et al.*, 2002).

In vitro stimulated splenic B cells from old mice are deficient in CSR and secondary Ig production, largely due to decreased induction of E47 and AID (Frasca *et al.*, 2004b; Frasca *et al.*, 2008a; Frasca and Blomberg, 2009). Both E47 protein and mRNA levels are decreased after stimulation of splenic B cells from old mice as compared with young mice. As E47 was decreased in splenic B cells from old mice, we investigated whether AID might also be lower in these cells. We found that unstimulated cells from both young and old mice express indiscernible levels of AID-specific mRNA. Stimulation of B cells from young mice induced an increase in mRNA expression at days three and four whereas optimal mRNA expression was attained at day five. Stimulation of B cells from old mice induced AID mRNA expression at days four and five, but the level of expression was lower (about five-fold) as compared to that exhibited by B cells from young Balb/c mice (Frasca *et al.*, 2004b; Table 9.1). E47 and AID were decreased to the same extent in old *versus* young B cells in response to different stimuli, such as anti-CD40/IL-4 (Frasca *et al.*, 2004b), BAFF/IL-4 (Frasca *et al.*, 2007b) or LPS (Frasca and Blomberg, unpublished data). As a consequence of the decrease in AID levels in old *versus* young B cells, less class switch DNA products were obtained five days after stimulation, whereas germline transcripts were indistinguishable in young and old B cells. These data are consistent with the defect in old B cells occurring at the CSR event and not due to problems with accessibility, or with the cytokine signaling pathway which leads to DNA accessibility. Decreases in E47 with age, in addition to directly decreasing levels of AID, may also directly affect SHM of the Ig loci by decreasing Ig gene transcription (Michael *et al.*, 2003) or by other mechanisms (Schoetz *et al.*, 2006).

Table 9.1. **Effects of age on mouse and human B cells**

Parameter measured	Mice (> 18 months*)	Humans (> 65 years)
PP2A^	↑ (3-fold) (submitted)	nd
p38 MAPK	= (Frasca 2005, 2007)	nd
phospho-p38MAPK	↓ (4-fold) (Frasca 2005, 2007)	nd
TTP	↑ (3-fold) (Frasca 2007)	nd
phospho-TTP	↓ (10-fold) (Frasca 2007)	nd
E47	↓ (5-fold) (Frasca 2003, 2004a,b)	↓ (3-fold) (Frasca 2008a)
AID	↓ (5-fold) (Frasca 2004b)	↓ (3/4-fold) (Frasca 2008a)
CSR	↓ (5-fold) (Frasca 2004b)	↓ (4-fold) (Frasca 2008a)
In vitro IgG	↓ (5/6-fold) (Frasca 2004b)	↓ (4/5-fold) (Frasca 2008a)

*Balb/c mice. ^Abbreviations are described in the text. nd, not done.

In order to better demonstrate the direct connection between E47, AID and CSR, we performed experiments with young and old E2A$^{+/-}$ mice. In both young and old E2A$^{+/-}$ mice the reduction in CSR to IgG1, IgG2a, IgG3 and IgE was about half, as compared to wild-type controls. E2A$^{+/-}$ mice and their wild-type controls are C57BL/6 mice. The difference in isotype production by splenic B cells from young and old C57BL/6 mice is less than that seen with B cells from young and old BALB/c mice, and may be explained as a difference between the two mouse strains. Moreover, the amount of switched isotypes released in cultures of splenic B cells from young E2A$^{+/-}$ mice is comparable to that of old wild-type mice. When AID was measured in splenic B cells from E2A$^{+/-}$ and wild-type young and old mice, it was found to be lower in E2A$^{+/-}$ mice, i.e. in the young E2A$^{+/-}$ as compared to young wild-type and in old E2A$^{+/-}$ mice as compared to old wild-type, as well as reduced in old wild-type as compared to young wild-type (Frasca *et al.*, 2004b).

These results together demonstrate that aging downregulates E47 mRNA and protein levels, which in turns causes less AID, leading to less Ig class switch, which affects the quality of antibody produced.

9.2.1 *Molecular mechanisms for reduced CSR in aging*

9.2.1.1 *E47 mRNA stability is decreased in aging*

RNA stability assays have indicated that the rate of E47 mRNA decay is accelerated in splenic B cells from old mice but E47 protein degradation rates are comparable in young *versus* old B cells, indicating that the regulation of E47 expression in activated old splenic B cells occurs primarily by mRNA stability (Frasca *et al.*, 2005). In contrast to splenic-activated B cells, E47 mRNA expression is comparable in bone marrow-derived IL-7-expanded pro-B/early pre-B cells from young and old mice (van der Put *et al.*, 2004). The reduced expression and DNA-binding of the E12/E47 transcription factor in aged B cell precursors is due to reduced protein stability (van der Put *et al.*, 2004; King *et al.*, 2007).

The stability of labile mRNA (like E47 in splenic B cells from old mice) may be controlled by signal transduction cascades, where the final product of the cascade phosphorylates a protein that interacts with adenylate/uridylate-rich elements (ARE) in the 3′ untranslated region (UTR) of mRNA and modifies its stability (Chen and Shyu, 1995). ARE sequences have been found in the 3′ UTR of many mRNAs, including E47. The E47 mRNA is a class I/III, because it contains pentamer AUUUA in the genomic poly-A end of the 3′ UTR plus several U-rich and AU-rich elements. At least part of the decreased stability of E47 mRNA seen in aged B cells is mediated by proteins, in particular by tristetraprolin (TTP; Frasca *et al.*, 2007a), a physiological regulator of mRNA stability (Taylor *et al.*, 1996; Phillips *et al.*, 2004). TTP-deficient mice develop a severe inflammatory syndrome, including polyarticular arthritis, myeloid hyperplasia, autoimmunity and cachexia (Taylor *et al.*, 1996), which is largely due to the increased stability of mRNAs for TNF-α, and the resulting enhanced secretion of proinflammatory cytokines (Taylor *et al.*, 1996; Phillips *et al.*, 2004). TTP binds the 3′ UTR of its own mRNA, which also contains AREs and stimulates deadenylation which leads to mRNA degradation (Brooks *et al.*, 2004; Tchen *et al.*, 2004). The TTP protein is a low-abundance cytosolic protein whose levels are dramatically induced by lipopolysaccharide (LPS). The protein is stable once induced, in contrast with its labile mRNA (Cao *et al.*, 2004; Brook *et al.*, 2006; Hitti *et al.*, 2006).

9.2.1.2 *TTP is increased in aging and decreases E47 mRNA stability*

We have shown that TTP is induced by LPS stimulation in purified splenic murine B cells and that TTP mRNA and protein expression are higher in old than in young B cells (Frasca *et al.*, 2007a). TTP is also induced by other stimuli, such as LPS/IL-4, anti-CD40, anti-CD40/IL-4 or IL-4. In all cases, the levels of TTP in old B cells were significantly higher as compared with those in young B cells. LPS was the best stimulus, but LPS/IL-4 and anti-CD40/IL-4 are also good stimuli to induce TTP. The different levels of expression of TTP in young and old B cells stimulated with LPS or with anti-CD40/IL-4 are consistent with different E47 mRNA expression in young and old B cells observed before (Frasca *et al.*, 2004b).

In order to demonstrate that TTP is involved in the degradation of the E47 mRNA, we performed a series of experiments in which the mRNA from stimulated young B cells was incubated with cytoplasmic extracts of activated B cells from young and old mice. Subsequently, E47 mRNA levels were revealed by real-time PCR. Results clearly indicate that cytoplasmic lysates from old B cells induced more degradation of E47 mRNA than those from young B cells. When TTP was removed from both the young and old cytoplasmic lysates by immunoprecipitation with specific anti-TTP antibody (IP) before the interaction with the mRNA occurs, the expression of E47 mRNA was significantly increased in both cases, but increased much more when TTP was removed from the old lysates (Frasca *et al.*, 2007a). In these experiments, poly(A) minus RNAs (including small RNAs) were removed, and only mRNA was present in the mixture with proteins. Thus, at present, we do not know the contribution of microRNAs (miRNAs) to the degradation of E47 mRNA. Retroviral rescue of old B cell function (CSR, AID) was accomplished with E47 without the 3′ UTR (Landin and Blomberg, unpublished data).

Both TTP protein expression and function in B cells are regulated by p38 MAPK and there is less phospho-p38 in old activated B cells (Frasca *et al.*, 2007a; Table 9.1). Also there is less phosphorylated TTP in old- than in young-activated splenic B cells which leads to more binding of TTP to the 3′ UTR of E47 mRNA, therefore decreasing its stability. Our studies demonstrate for the first time that TTP is regulated in activated B cells during aging, that TTP is involved in the degradation of the E47 mRNA and show the molecular mechanism for the decreased expression of E47, AID and CSR in aged murine B cells.

9.2.1.3 *PP2A is increased in aging and reduces TTP and MAPK activity and E47 mRNA stability*

Our recent results (Frasca and Blomberg, unpublished data) show that PP2A, a serine/threonine protein phosphatase that plays an important role in the regulation of several major signaling pathways, is increased in old B cells. Also, PP2A phosphatase activity is increased in old B cells. As a consequence of this, higher phosphatase activity in old B cells, p38 MAPK is less phosphorylated as compared with young B cells. PP2A dephosphorylation of either p38 MAPK and/or TTP may account for more binding of the hypophosphorylated TTP to the 3′ UTR of the E47 mRNA, inducing its degradation. Our data supports this mechanism for regulating the molecular pathways that lead to reduced antibody responses in aging. The full elucidation of these molecular pathways will help to design effective strategies to prevent age-associated pathologies, potentiate the response to exogenous antigens and improve the biological quality of life in aging individuals.

9.2.1.4 *Age-related differences in human B cell responses*

We have extended our studies in mice to investigate whether aging also affects antibody production and both E47 and AID expression in B cells isolated from the peripheral blood of human subjects. Our results indicate that elderly humans have reduced percentages of $CD19^+$ total B cells and switched memory B cells, and increased percentages of naïve B cells. This decrease in switched memory B cells and the increase in the percentage of naïve B cells is consistent with the intrinsic defect in the ability of old B cells to undergo CSR that we also see *in vitro*. We have also found less IgG production in *in vitro* stimulated B cells with age due to less AID and E47 (Frasca *et al.*, 2008a). In our experiments, $CD19^+$ B cells, isolated from the peripheral blood of subjects of different ages, were stimulated *in vitro* with either anti-CD40/IL-4 (Frasca *et al.*, 2008a) or CpG (Frasca and Blomberg, unpublished data). CSR was measured by IgG gamma (γ) circle transcripts (CTs) which were reduced in B cells from elderly individuals as compared to the younger ones. We have also investigated germline transcripts and found that they are present in B cells from young and elderly individuals in similar amounts, as we have seen in the murine system (Frasca *et al.*, 2004b). Thus, our results again show less CSR products, here in humans, and suggest that the defect in aged B cells occurs at the level of CSR and is not due to problems with accessibility, nor with the cytokine signaling pathway, as we have also previously shown in murine B cells (Frasca *et al.*, 2004b). These results were also continued by measuring IgG production by ELISA of culture supernatants after seven days of stimulation with anti-CD40/IL-4 or CpG where we found that IgG levels in culture supernatants decreased with age and, as expected, parallel the data of γ CTs.

Because the reduction in CSR in B cells from elderly individuals could be due to reduced levels of AID, as already shown in murine B cells, we investigated the levels of AID expression in anti-CD40/IL-4- or CpG-stimulated $CD19^+$ B cells from individuals of different ages. AID mRNA levels measured by qPCR were lower in B cells from old as compared to young subjects.

To gain insight into the mechanisms underlying AID regulation, we analyzed the levels of E47 expression in B cells from young and elderly individuals. We predicted from the mouse studies that E47 expression levels would be lower in B cells from old as compared to young subjects and this was confirmed here in humans. These results demonstrate that both E47 and AID, progressively decrease with age in a similar manner and we found a significant correlation between E47 and AID ($r = 0.80$, $p<0.01$). Higher E47 was also associated with higher CTs ($r = 0.65$, $p<0.01$). Likewise, there was a positive significant association between AID and CTs where greater AID was associated with greater CTs ($r = 0.79$, $p<0.01$; Table 9.1).

We have formally ruled out the hypothesis that our results on reduced CSR with age are due to an age-related reduction in the numbers of memory B cells, and therefore are due to an intrinsic defect in the memory B cells. This hypothesis was less likely because we have shown that naïve and IgM memory were either increased or unchanged (percentage wise) with age (Frasca *et al.*, 2008a) and we showed in a separate experiment that the IgM memory cells ($CD19^+CD27^+IgG^-IgA^-$) are the memory cells which class switch *in vitro* (Frasca *et al.*, 2008a). To formally test the alternate hypothesis, we sorted naïve ($CD19^+CD27^-$) and memory ($CD19^+CD27^+$, including IgM memory cells which are IgG^- IgA^-) B cells, which were stimulated *in vitro* by anti-CD40/IL-4 and $F(ab')_2$ fragments of anti-human IgM, used as surrogate antigen, because naïve B cells require the activation of the BCR signal transduction to undergo class switch (Bernasconi *et al.*, 2002). These results indicate that there is an intrinsic defect in the ability of naïve and IgM memory cells to class switch and therefore the defect we observed in the ability of $CD19^+$ cells to class switch upon *in vitro* stimulation with anti-CD40/IL-4 does not simply depend on a reduction in the numbers of memory B cells (Frasca *et al.*, 2008a). It also emphasizes the dual defect in aged B cells: there are less of them and their function (in ability to class switch) is also less on a per cell basis.

We recently started to establish biological significance of these biomarkers *in vivo* and initiated a series of experiments to measure the antibody response to influenza vaccination by hemagglutination inhibition assay (HI) and associated this with the B cell response to these antigens *in vitro* (as read out by increase in AID). Infectious pathogenic diseases are fairly common, with influenza alone affecting up to an estimated 50 million people each year in the US. Approximately 40,000 deaths are attributed to influenza, with most of the affected being the elderly and those patients with weak immune systems. We recently started investigating the specific response of B cells from subjects of different ages to influenza vaccination. Our preliminary results (Fig. 9.1) show that the *in vitro* AID response of B cells to vaccination and the serum HI response to vaccination are correlated and that both are

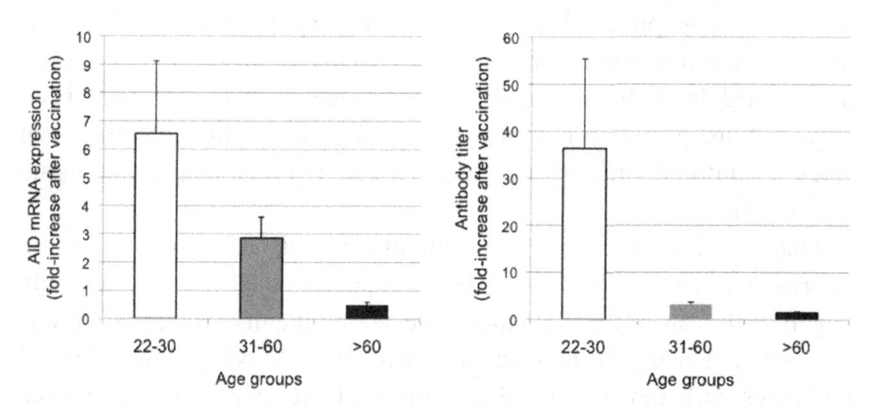

Figure 9.1. **The *in vitro* response and the serum response to vaccination are decreased with age.** B cells were isolated from the peripheral blood of subjects of different ages at t = 0 (before vaccination) and at t = 28 (one month after vaccination) and cultures for seven days with the flu vaccine Fluarix or Fluvirin, with which they were immunized. The response to either vaccine was equivalent. AID was measured by qPCR. Sera were collected from the same subjects at t = 0 and t = 28 and analyzed by HI (hemagglutination inhibition/antibody titer) to evaluate antibody production to the vaccine. Group 22–30: 6 subjects; group 31–60: 5 subjects; group >60: 5 subjects. Data are from the 2008 flu season.

decreased with age. We are currently pursuing these studies with more subjects and with both seasonal as well as H1N1 vaccine response. These results indicate that these biomarkers would more accurately track immune responses and activity than markers currently used, and would be better indicators of vaccine and therapeutic agent effectiveness in humans.

9.3 AID in Autoimmunity

There is compelling evidence that B cells play a critical role in the pathogenesis of many of the organ-specific and systemic autoimmune diseases. Systemic autoimmune disorders with a known B cell component include systemic lupus erythematosus (SLE), rheumatoid arthritis, mixed connective tissue disorder and scleroderma, among others. Many of the hallmark antibodies associated with these disorders are against nuclear components such as DNA, ribonuclear proteins and DNA repair proteins, antigens not readily exposed to the cells of the immune system. It is not understood why antinuclear antibodies develop but defective apoptosis, defective clearance of apoptotic material, excess necrosis from ongoing chronic damage, defects in toll receptor signaling and a breakdown in B and T cell tolerance mechanisms have been suggested as potential culprits. Ultimately, these autoantibodies can trigger an autoimmune cascade characterized by inflammation and tissue destruction.

One way B cells contribute to autoimmune disorders is by secreting autoreactive antibodies. However, it appears this is not the only way B cells can trigger or exacerbate autoimmunity. Recently, it was demonstrated that autoimmune mice with B cells but lacking secreted antibodies, still develop a milder form of lupus nephritis, while mice completely lacking B cells did not (Chan *et al.*, 1999). This suggested an antibody-independent contribution of B cells to autoimmunity. Subsequent studies provided evidence that B cells contribute to autoimmunity by activating autoreactive $CD4^+$ T cells, likely as antigen-presenting cells. Therefore, B cells contribute to autoimmune disease through the production of autoreactive pathogenic antibodies, and as antigen-presenting cells to autoreactive T cells. Given the importance of antibody and B cell receptor (BCR) specificity to self-antigens in autoimmune disorders, it is of great significance to delineate the differential contribution of the various mechanisms that contribute to diversity, specificity and function of the BCR and of antibodies. The three main mechanisms responsible for generating diversity and increasing specificity of the B cell repertoire are V(D)J recombination – which generates the pre-immune naïve repertoire – and the antigen-driven processes, SHM and CSR. AID is critical to two of these mechanisms: SHM and CSR. It is through its requirement for both of these processes that AID plays a major role in B cell-mediated autoimmunity (Fig. 9.2).

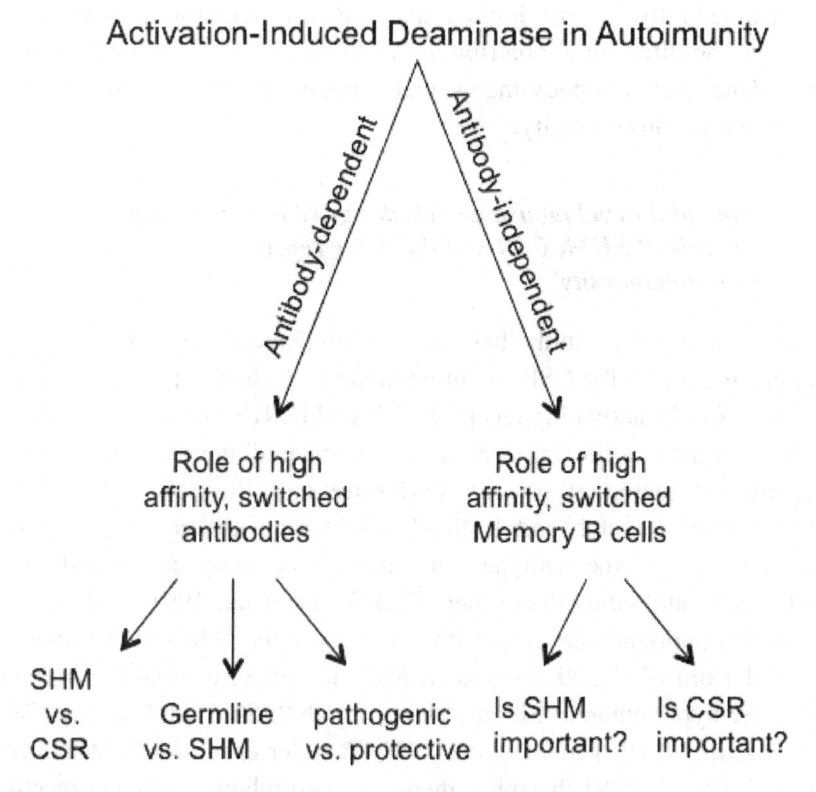

Figure 9.2. **The two main pathways of potential AID contribution to B cell mediated autoimmunity.**

AID triggers SHM and CSR during secondary B cell responses by deaminating cytosines in the variable and switch regions of the immunoglobulin locus (Rada *et al.*, 2002). It is absolutely required for both of these reactions: B cells from organisms lacking AID activity can only secrete unmutated, germline IgM antibodies (Muramatsu *et al.*, 2000; Revy *et al.*, 2000). Therefore, mouse models wherein AID is lacking in an autoimmune background are powerful tools to examine the contribution of autoreactive germline antibodies to autoimmunity. Isolating the contribution of SHM *vs.* CSR to autoimmunity is more difficult since no molecule has been described to date whose deficiency eliminates one of the mechanisms and not the other. However, through genetic manipulation of AID expression in various backgrounds, we have generated a series of mice that either lack SHM, CSR or are characterized by a very limited naïve B cell repertoire, in order to examine the differential contribution of these processes to the antibody-dependent and antibody-independent pathways by which B cells contribute to autoimmunity.

9.3.1 *Potential novel mouse models designed to distinguish the role of SHM, CSR and the naïve repertoire in autoimmunity*

Often, pathogenic antibodies in autoimmunity are IgG, strongly implicating a role for CSR in autoimmunity. Consistent with this, mice deficient for the activating receptors FcR and FcRIII experienced reduced kidney damage while mice deficient in the inhibitory receptor FcRII, experienced increased severity (Sylvestre and Ravetch, 1994; Takai *et al.*, 1996; Ravetch and Bolland, 2001). In addition, there is good evidence that some isotypes are more frequently associated with pathogenic antibodies than others (Takahashi *et al.*, 1991). SHM also plays an important role in systemic autoimmunity. Most autoantibodies derived from SLE patients and in MRL/lpr mice, a mouse model for SLE, are hypermutated (Shlomchik *et al.*, 1990; Shlomchik *et al.*, 1987; Radic *et al.*, 1989; van Es *et al.*, 1991; Winkler *et al.*, 1992; Wellmann *et al.*, 2005). In MRL/lpr mice, there was a correlation with autoreactive antibodies and specific mutations, particularly those introducing arginines into the CDRs (Radic *et al.*, 1989). It is likely that because SHM is random in relation to affinity, new mutations could be introduced that increase affinity to self-antigens generating autoreactive antibodies. In a normal individual, these mutated B cells bearing autoreactive BCR's and capable of secreting autoreactive antibodies, are eliminated. However, in autoimmune-prone individuals peripheral tolerance checkpoints may be defective, resulting in the recruitment of these cells into the memory compartment. Such disruption can be accomplished by, among others, defects in B or T cell signaling in germinal centers, defective apoptosis or even disruption of the germinal center environment such that SHM can occur outside of germinal centers without proper monitoring of newly-generated autoreactive B cells (William *et al.*, 2002). Finally, the naïve repertoire also contributes to autoimmunity. Newly-generated B cells, each bearing a different and unique BCR, may harbor autoreactivity-enhancing amino acids in their CDR3s formed during V(D)J recombination. This is particularly true for large, positively charged amino acids such as arginines, which can be either encoded in the D or J elements or formed during N-region addition by terminal deoxynucleotide transferase or TDT. These cells are normally modified by receptor editing or deleted during central tolerance in the bone marrow. In most autoimmune mice, however, these cells readily survive into the periphery and may become the precursors to high-affinity autoreactive B cells. Consistent with the formation of autoreactive receptors during V(D)J recombination is the finding that mice deficient in TDT experienced a significant decrease in glomerulonephitis (Molano *et al.*, 2003; Robey *et al.*, 2004).

In order to distinguish the role of these various mechanisms that contribute to antibody diversity and versatility, we crossed various strains with B cell defects into the AID-deficient MRL/lpr background. For example, muS-/-.MRL/lpr mice (with a defect in the exon encoding the secretory domain for IgM) have a full naïve repertoire, can undergo SHM and CSR, but cannot secrete IgM (Boes *et al.*, 2000). They also have elevated levels of IgG autoantibodies to dsDNA, increased deposition of immune complexes in glomeruli, severe glomerulonephritis and early onset of lupus-like syndrome. Crossing the strain to AID-deficient MRL/lpr resulted in mice lacking any secreted antibodies, or

SHM, but still maintaining a full naïve repertoire of B cells with surface expression of IgM. These mice (lacking any secreted antibodies) can be compared to AID.MRL/lpr mice (secreting only IgM) to isolate the impact of secreted IgM in the lupus syndrome of MRL/lpr mice.

JHT.MRL/lpr mice completely lack B cells and their products and are similar to mice generated by Shlomchik and colleagues (Chan *et al.*, 1999). These mice have a very mild manifestation of the MRL/lpr syndrome. Rescuing B cells in these mice by introducing a transgene-encoding membrane IgM but incapable of secreting antibodies (JHT.mIg.MRL/lpr), restores some of the phenotype, in particular the infiltration of mononuclear cells into the kidneys and other tissues, suggesting an antibody-independent pathway for B cells in autoimmunity. However, these mice express a single V_H (186.2), which is not normally autoreactive. This led us to speculate that the relevant B cells became autoreactive through SHM and affinity maturation against self-antigen. We are testing this hypothesis by crossing the JHT.mIg.MRL/lpr to AID.MRL/lpr mice. The resulting offspring lack secreted antibodies, have B cells from a limited repertoire and lack SHM. These mice can be used to compare to AID wild-type JHT.mIg.MRL/lpr, and will provide clues to the significance of SHM and affinity maturation to the antibody-independent contribution of B cells to autoimmunity.

Another possible strategy to isolate the role of CSR from SHM, is by altering the amino-acid sequence of the AID protein through the generation of specific AID-knocked-in mice. Loss of the C-terminal nuclear export signal region of AID results in normal SHM but impaired CSR (Ta *et al.*, 2003; Barreto *et al.*, 2003). This possibility, however, needs more clarification since it is unclear if AID-mediated mutation in these mice is targeted to Ig loci, considering that loss of the NES is expected to increase AID levels in the nucleus (Brar *et al.*, 2004), which may cause widespread AID-mediated deamination.

9.3.2 *AID-deficient autoimmune-prone mice*

To generate the novel mouse models described above, our laboratory crossed AID deficiency into the autoimmune-prone MRL/lpr strain

(Jiang *et al.*, 2007). This is the first model combining autoimmunity with AID-deficiency. MRL/lpr mice develop a lupus-like syndrome that shares many similarities with human SLE such as high levels of circulating autoantibodies, particularly to double-stranded DNA (dsDNA), deposition of immune complexes, particularly in the kidneys, and glomerulonephritis (Theofilopoulos and Dixon, 1985). When AID-deficient, these mice experience a highly significant decrease in kidney damage, and a dramatic increase in survival to levels that exceeded even mice lacking secreted antibodies, suggesting that improved survival did not result just from the lack of pathogenic IgG antibodies but that other factors also contributed. The proportion of activated autoreactive T cells were unchanged suggesting that the increase in survival did not come from the antibody-independent role of B cells in autoimmunity. These mice have very high levels of autoreactive IgM, not only demonstrating that autoreactive IgM is not pathogenic in lupus nephritis but also suggesting that these IgM antibodies may be protective. We have generated hybridomas secreting anti-dsDNA IgM and preliminary data suggests these antibodies may in fact protect MRL/lpr mice from lupus nephritis (Jiang and Diaz, unpublished data). Interestingly, autoimmune mice lacking secreted IgM but with secreted IgG experienced accelerated autoimmunity (Boes *et al.*, 2000).

AID heterozygous MRL/lpr mice experienced a delay in the onset of lupus nephritis (Jiang *et al.*, 2009). AID haploinsufficiency was a surprising finding because it is well known that only a fraction of cellular AID reaches the nucleus where its activity is required (Brar *et al.*, 2004). While a CSR defect was detectable in *in vitro* switch assays, serum levels of switched antibodies were comparable to controls, even in very young asymptomatic mice. These mice had very low levels of high-affinity anti-dsDNA IgG antibodies likely to be from reduced SHM, strongly suggesting that the delay in the onset of autoimmunity was due to impaired affinity maturation of B cells against self-antigens. These data suggest a role for SHM and affinity maturation that is independent of CSR in the generation of pathogenic autoantibodies.

9.3.3 *AID overexpression effects and autoimmunity in mice*

Using a different mouse model of autoimmune disease, Mount and colleagues found that BXD2 mice have overexpression of AID in B cells, and that this correlated with high levels of pathogenic IgG antibodies that harbored higher than average numbers of mutations (Hsu *et al.*, 2007). The authors concluded that many of these antibodies acquired mutations that increased autoreactivity of the BCR and antibodies, contributing to autoimmunity. Interestingly, inhibition of $CD4^{+}CD28^{+}$ T cells in these mice resulted in normalization of AID levels in B cells and a drop in the levels of pathogenic antibodies. Studies with transgenic mice with overexpression of AID limited to B cells revealed a negative regulatory mechanism that inactivates AID protein (Muto *et al.*, 2006). Perhaps, AID levels are tightly monitored in B cells, not only to prevent deregulated deamination that can lead to neoplasia in activated B cells, but also to prevent the generation of autoantibodies through high levels of SHM. In fact most of the autoantibodies isolated from SLE patients are mutated and some of these mutations enhance recognition to self-antigens (van Es *et al.*, 1991).

Recently, it was shown that AID expression can be activated in B cells with exposure to estrogen, due to a specific interaction of the estrogen complex with the promoter region of AID (Pauklin *et al.*, 2009). This expression was not limited to B cells but could also be seen in breast- and ovary-derived tissues. A connection between AID levels, autoimmunity and estrogen may provide the basis for the higher incidence of autoimmune disorders in women (Pauklin *et al.*, 2009; Maul and Gearhart, 2009), although the mechanism wherein estrogen-mediated activation of AID may induce autoimmunity is not understood. It is possible that the increase in genomic instability from increased AID expression may contribute to autoimmunity. It is well known that agents inducing DNA damage such as UV radiation and bleomycin can exacerbate SLE. It is also possible that hyperactivation of AID increases the probability of generating autoreactive antibodies through SHM as seen with BXD2 mice. While interesting, a connection between estrogen-mediated activation of AID and autoimmunity remains speculative and awaits direct testing.

9.3.4 *AID deficiency and autoimmunity in humans*

Unlike its mouse counterpart, AID deficiency in humans is associated with an increase in the incidence of certain autoimmune disorders, particularly the immune cytopenias, such as autoimmune hemolytic anemia, autoimmune thrombocytopenia or autoimmune neutropenia (Quartier *et al.*, 2004). Interestingly, these autoimmunities are also associated with other primary immunodeficiencies such as common variable immune deficiency, and rarely, severe combined immunodeficiency (Arkwright *et al.*, 2002). The common denominator appears to be the inability of the immunodeficient individual to clear persisting pathogens (Arkwright *et al.*, 2002). It is possible that in those individuals with a residual lymphocyte population, there is compensation by relaxing tolerance checkpoints in order to deal with the pressure from persisting infection, allowing expansion and activation of autoreactive lymphocytes that would otherwise be prevented from participating in the immune response. It was also suggested that in these individuals, compensatory inflammatory responses that poorly discriminate infected cells from healthy cells, contribute to the immune-mediated tissue destruction (Arkwright *et al.*, 2002). In AID-deficient patients, it is possible that the high levels of IgM may be associated with development of autoimmune disease from autoreactive IgM. Indeed, IgM antibodies against erythrocytes have been associated with autoimmune hemolytic anemia in patients, and transgenic mice expressing an anti-erythrocyte IgM specificity develop a similar syndrome (Sokol *et al.*, 1998; Okamoto *et al.*, 1992). However, we did not detect evidence of an increase in immune cytopenias in AID-deficient MRL/lpr mice, in spite of these mice having very high levels of autoreactive IgM. This discordance with the human data may be explained by the fact that mice in specific pathogen-free facilities may experience relative low exposure to pathogens, but this remains to be tested. Deciphering the reasons for this discrepancy in AID-deficient people and in mice will likely yield novel insights into the role of IgM versus IgG in the pathogenesis of autoimmune disorders and into the relationship between autoimmunity, infection and immunodeficiency.

9.4 Conclusion

It is a fact that a significant way by which AID contributes to autoimmune disease is through the development of high-affinity, isotype-switched autoantibodies. Less clear is its contribution to the antibody-independent mechanism by which B cells cause autoimmunity. At this point, it is unknown whether affinity to self-antigen or the isotype of the BCR matters to this aspect of B-cell mediated autoimmunity. Our above described novel mouse models using AID deficiency in combinations with other B cell defects in MRL/lpr mice will likely generate useful information in this regard.

While both SHM and CSR play an important role in the antibody-dependent role of B cell mediated autoimmunity, their differential contributions remain difficult to assess, mostly because no molecule has been identified that impairs one mechanism and not the other that could be rendered deficient in an autoimmune background. This is an important question because the role each of these mechanisms play in particular autoimmune diseases is likely to reveal important aspects of the B-cell population secreting the pathogenic antibodies. For example, in myasthenia gravis, SHM and affinity maturation may play a critical role in generating high-affinity antibodies against the acetylcholine receptor (Sims *et al.*, 2001), suggesting the involvement of memory B cells, while antibodies against collagen type II which are associated with collagen-induced arthritis are often in germline configuration but can be isotype-switched to IgG (Mo *et al.*, 1994). In addition, if affinity maturation through SHM plays an important role in the generation of pathogenic antibodies, it suggests not only a breakdown in peripheral tolerance in autoimmune patients but that self-antigen can be the selected factor driving the germinal center reaction.

Interestingly, AID levels appear to have a profound impact in autoimmunity. AID-heterozygous MRL/lpr mice experienced a delay in the onset of lupus nephritis that best correlated with dramatically reduced levels of high-affinity anti-dsDNA IgG antibodies (Jiang *et al.*, 2009). These results suggest that reducing the levels of AID in B cells could be a potential novel therapy in autoimmune patients. Indeed, normalization of AID levels in BXD2 mice, correlated with a decrease in pathogenic antibodies. However, the association of increased autoimmunity with AID deficiency in humans, and with immunodeficiency in aged individuals, highlights the difficulty in translating mouse studies for clinical applications. This is compounded by the fact that not all pathogenic antibodies in autoimmunity are mutated or isotype-switched. A more rational approach will likely require disease-specific strategies based on the culprit B cell population (naïve, memory, etc.) and which types of antibodies directly contribute to tissue damage (IgM, IgG, mutated or germline).

9.5 Acknowledgements

This work was in part supported by grants NIH AG23717 (B.B.B.), NIH AG28586 (B.B.B.) and NIH AG32576 (B.B.B.) and by the Intramural Research Program of the National Institutes of Health, National Institute of Environmental Health Sciences (M.D.).

9.6 References

1. Albright J.F., Albright J.W. (2003) *Aging, Immunity, and Infection*, Totowa, NJ, Humana Press.
2. Arkwright P.D., Abinun M., Cant A.J. (2002) Autoimmunity in human primary immunodeficiency diseases. *Blood* **99**: 2694–2702.
3. Barreto V., Reina-San-Martin B., Ramiro A.R. *et al.* (2003) C-terminal deletion of AID uncouples class switch recombination from somatic hypermutation and gene conversion. *Mol Cell* **12**: 501–508.
4. Basu U., Chaudhuri J., Alpert C. *et al.* (2005) The AID antibody diversification enzyme is regulated by protein kinase A phosphorylation. *Nature* **438**: 508–511.
5. Bernasconi N.L., Traggiai E., Lanzavecchia A. (2002) Maintenance of serological memory by polyclonal activation of human memory B cells. *Science* **298**: 2199–2202.
6. Boes M., Schmidt T., Linkemann K. *et al.* (2000) Accelerated development of IgG autoantibodies and autoimmune disease in the absence of secreted IgM. *Proc Natl Acad Sci USA* **97**: 1184–1189.
7. Brar S.S., Watson M., Diaz M. (2004) Activation-induced cytosine deaminase (AID) is actively exported out of the nucleus but retained by the induction of DNA breaks. *J Biol Chem* **279**: 26395–26401.
8. Brook M., Tchen C.R., Santalucia T. *et al.* (2006) Posttranslational regulation of tristetraprolin subcellular localization and protein stability by p38 mitogen-activated

protein kinase and extracellular signal-regulated kinase pathways. *Mol Cell Biol* **26**: 2408–2418.

9. Brooks S.A., Connolly J.E., Rigby W.F. (2004) The role of mRNA turnover in the regulation of tristetraprolin expression: evidence for an extracellular signal-regulated kinase-specific, AU-rich element-dependent, autoregulatory pathway. *J Immunol* **172**: 7263–7271.
10. Cancro M.P., Hao Y., Scholz J.L. *et al.* (2009) B cells and aging: molecules and mechanisms. *Trends Immunol* **30**: 313–318.
11. Cao H., Tuttle J.S., Blackshear P.J. (2004) Immunological characterization of tristetraprolin as a low abundance, inducible, stable cytosolic protein. *J Biol Chem* **279**: 21489–21499.
12. Chan O.T., Hannum L.G., Haberman A.M. *et al.* (1999) A novel mouse with B cells but lacking serum antibody reveals an antibody-independent role for B cells in murine lupus. *J Exp Med* **189**: 1639–1648.
13. Chen C.Y., Shyu A.B. (1995) AU-rich elements: characterization and importance in mRNA degradation. *Trends Biochem Sci* **20**: 465–470.
14. Chong Y., Ikematsu H., Yamaji K. *et al.* (2005) CD27(+) (memory) B cell decrease and apoptosis-resistant CD27(-) (naive) B cell increase in aged humans: implications for age-related peripheral B cell developmental disturbances. *Int Immunol* **17**: 383–390.
15. Connoy A.C., Trader M., High K.P. (2006) Age-related changes in cell surface and senescence markers in the spleen of DBA/2 mice: a flow cytometric analysis. *Exp Gerontol* **41**: 225–229.
16. Dorsett Y., McBride K.M., Jankovic M. *et al.* (2008) MicroRNA-155 suppresses activation-induced cytidine deaminase-mediated Myc-Igh translocation. *Immunity* **28**: 630–638.
17. Dorsett Y., Robbiani D.F., Jankovic M. *et al.* (2007) A role for AID in chromosome translocations between c-myc and the IgH variable region. *J Exp Med* **204**: 2225–2232.
18. Durandy A. (2002) Hyper-IgM syndromes: a model for studying the regulation of class switch recombination and somatic hypermutation generation. *Biochem Soc Trans* **30**: 815–818.
19. Engwerda C.R., Handwerger B.S., Fox B.S. (1994) Aged T cells are hyporesponsive to costimulation mediated by CD28. *J Immunol* **152**: 3740–3747.
20. Ephrussi A., Church G.M., Tonegawa S. *et al.* (1985) B lineage-specific interactions of an immunoglobulin enhancer with cellular factors in vivo. *Science* **227**: 134–140.
21. Frasca D., Blomberg B.B. (2009) Effects of aging on B cell function. *Curr Opin Immunol* **21**: 425–430.
22. Frasca D., Landin A.M., Alvarez J.P. *et al.* (2007a) Tristetraprolin, a negative regulator of mRNA stability, is increased in old B cells and is involved in the degradation of E47 mRNA. *J Immunol* **179**: 918–927.
23. Frasca D., Landin A.M., Lechner S.C. *et al.* (2008a) Aging down-regulates the transcription factor E2A, activation-induced cytidine deaminase, and Ig class switch in human B cells. *J Immunol* **180**: 5283–5290.

24. Frasca D., Landin A.M., Riley R.L. *et al.* (2008b) Mechanisms for decreased function of B cells in aged mice and humans. *J Immunol* **180**: 2741–2746.
25. Frasca D., Nguyen D., Riley R.L. *et al.* (2003) Decreased E12 and/or E47 transcription factor activity in the bone marrow as well as in the spleen of aged mice. *J Immunol* **170**: 719–726.
26. Frasca D., Riley R.L., Blomberg B.B. (2007b) Aging murine B cells have decreased class switch induced by anti-CD40 or BAFF. *Exp Gerontol* **42**: 192–203.
27. Frasca D., van Der Put E., Landin A.M. *et al.* (2005) RNA stability of the E2A-encoded transcription factor E47 is lower in splenic activated B cells from aged mice. *J Immunol* **175**: 6633–6644.
28. Frasca D., van Der Put E., Riley R.L. *et al.* (2004a) Age-related differences in the E2A-encoded transcription factor E47 in bone marrow-derived B cell precursors and in splenic B cells. *Exp Gerontol* **39**: 481–489.
29. Frasca D., van Der Put E., Riley R.L. *et al.* (2004b) Reduced Ig class switch in aged mice correlates with decreased E47 and activation-induced cytidine deaminase. *J Immunol* **172**: 2155–2162.
30. Garcia G.G., Miller R.A. (2002) Age-dependent defects in TCR-triggered cytoskeletal rearrangement in CD4+ T cells. *J Immunol* **169**: 5021–5027.
31. Gardner E.M., Gonzalez E.W., Nogusa S. *et al.* (2006) Age-related changes in the immune response to influenza vaccination in a racially diverse, healthy elderly population. *Vaccine* **24**: 1609–1614.
32. Gibson K.L., Wu Y.C., Barnett Y. *et al.* (2009) B-cell diversity decreases in old age and is correlated with poor health status. *Aging Cell* **8**: 18–25.
33. Gonda H., Sugai M., Nambu Y. *et al.* (2003) The balance between Pax5 and Id2 activities is the key to AID gene expression. *J Exp Med* **198**: 1427–1437.
34. Han S., Yang K., Ozen Z. *et al.* (2003) Enhanced differentiation of splenic plasma cells but diminished long-lived high-affinity bone marrow plasma cells in aged mice. *J Immunol* **170**: 1267–1273.
35. Haynes L., Eaton S.M., Burns E.M. *et al.* (2003) CD4 T cell memory derived from young naive cells functions well into old age, but memory generated from aged naive cells functions poorly. *Proc Natl Acad Sci USA* **100**: 15053–15058.
36. Henthorn P., Kiledjian M., Kadesch T. (1990) Two distinct transcription factors that bind the immunoglobulin enhancer microE5/kappa 2 motif. *Science* **247**: 467–470.
37. Hitti E., Iakovleva T., Brook M. *et al.* (2006) Mitogen-activated protein kinase-activated protein kinase 2 regulates tumor necrosis factor mRNA stability and translation mainly by altering tristetraprolin expression, stability, and binding to adenine/uridine-rich element. *Mol Cell Biol* **26**: 2399–2407.
38. Hodes R.J. (1997) Aging and the immune system. *Immunol Rev* **160**: 5–8.
39. Hsu H.C., Wu Y., Yang P. *et al.* (2007) Overexpression of activation-induced cytidine deaminase in B cells is associated with production of highly pathogenic autoantibodies. *J Immunol* **178**: 5357–5365.
40. Jiang C., Foley J., Clayton N. *et al.* (2007) Abrogation of lupus nephritis in activation-induced deaminase-deficient MRL/lpr mice. *J Immunol* **178**: 7422–7431.

41. Jiang C., Zhao M.L., Diaz M. (2009) Activation-induced deaminase heterozygous MRL/lpr mice are delayed in the production of high-affinity pathogenic antibodies and in the development of lupus nephritis. *Immunology* **126**: 102–113.
42. Kee B.L., Bain G., Murre C. (2002) IL-7Ralpha and E47: independent pathways required for development of multipotent lymphoid progenitors. *EMBO J* **21**: 103–113.
43. King A.M., van Der Put E., Blomberg B.B. *et al.* (2007) Accelerated notch-dependent degradation of E47 proteins in aged B cell precursors is associated with increased ERK MAPK activation. *J Immunol* **178**: 3521–3529.
44. Klinman N.R., Kline G.H. (1997) The B-cell biology of aging. *Immunol Rev* **160**: 103–114.
45. LeMaoult J., Szabo P., Weksler M.E. (1997) Effect of age on humoral immunity, selection of the B-cell repertoire and B-cell development. *Immunol Rev* **160**: 115–126.
46. Linton P., Thoman M.L. (2001) T cell senescence. *Front Biosci* **6**: D248–261.
47. Linton P.J., Dorshkind K. (2004) Age-related changes in lymphocyte development and function. *Nat Immunol* **5**: 133–139.
48. Massari M.E., Murre C. (2000) Helix-loop-helix proteins: regulators of transcription in eucaryotic organisms. *Mol Cell Biol* **20**: 429–440.
49. Maul R.W., Gearhart P.J. (2009) Women, autoimmunity, and cancer: a dangerous liaison between estrogen and activation-induced deaminase? *J Exp Med* **206**: 11–13.
50. Michael N., Shen H.M., Longerich S. *et al.* (2003) The E box motif CAGGTG enhances somatic hypermutation without enhancing transcription. *Immunity* **19**: 235–242.
51. Miller C., Kelsoe G. (1995) Ig VH hypermutation is absent in the germinal centers of aged mice. *J Immunol* **155**: 3377–3384.
52. Mo J.A., Scheynius A., Nilsson S. *et al.* (1994) Germline-encoded IgG antibodies bind mouse cartilage in vivo: epitope- and idiotype-specific binding and inhibition. *Scand J Immunol* **39**: 122–130.
53. Molano I.D., Redmond S., Sekine H. *et al.* (2003) Effect of genetic deficiency of terminal deoxynucleotidyl transferase on autoantibody production and renal disease in MRL/lpr mice. *Clin Immunol* **107**: 186–197.
54. Muramatsu M., Kinoshita K., Fagarasan S. *et al.* (2000) Class switch recombination and hypermutation require activation-induced cytidine deaminase (AID), a potential RNA editing enzyme. *Cell* **102**: 553–563.
55. Muramatsu M., Nagaoka H., Shinkura R. *et al.* (2007) Discovery of activation-induced cytidine deaminase, the engraver of antibody memory. *Adv Immunol* **94**: 1–36.
56. Murre C., McCaw P.S., Baltimore D. (1989) A new DNA binding and dimerization motif in immunoglobulin enhancer binding, daughterless, MyoD, and myc proteins. *Cell* **56**: 777–783.
57. Muto T., Okazaki I.M., Yamada S. *et al.* (2006) Negative regulation of activation-induced cytidine deaminase in B cells. *Proc Natl Acad Sci USA* **103**: 2752–2757.

58. Notarangelo L.D., Lanzi G., Peron S. *et al.* (2006) Defects of class-switch recombination. *J Allergy Clin Immunol* **117**: 855–864.
59. Nussenzweig M.C., Alt F.W. (2004) Antibody diversity: one enzyme to rule them all. *Nat Med* **10**: 1304–1305.
60. Okamoto M., Murakami M., Shimizu A. *et al.* (1992) A transgenic model of autoimmune hemolytic anemia. *J Exp Med* **175**: 71–79.
61. Okazaki I.M., Kinoshita K., Muramatsu M. *et al.* (2002) The AID enzyme induces class switch recombination in fibroblasts. *Nature* **416**: 340–345.
62. Pasqualucci L., Kitaura Y., Gu H. *et al.* (2006) PKA-mediated phosphorylation regulates the function of activation-induced deaminase (AID) in B cells. *Proc Natl Acad Sci USA* **103**: 395–400.
63. Pasqualucci L., Migliazza A., Fracchiolla N. *et al.* (1998) BCL-6 mutations in normal germinal center B cells: evidence of somatic hypermutation acting outside Ig loci. *Proc Natl Acad Sci USA* **95**: 11816–11821.
64. Pauklin S., Sernandez I.V., Bachmann G. *et al.* (2009) Estrogen directly activates AID transcription and function. *J Exp Med* **206**: 99–111.
65. Pawelec G., Barnett Y., Forsey R. *et al.* (2002) T cells and aging, January 2002 update. *Front Biosci* **7**: d1056–1183.
66. Peres A., Bauer M., Da Cruz I.B. *et al.* (2003) Immunophenotyping and T-cell proliferative capacity in a healthy aged population. *Biogerontology* **4**: 289–296.
67. Phillips K., Kedersha N., Shen L. *et al.* (2004) Arthritis suppressor genes TIA-1 and TTP dampen the expression of tumor necrosis factor alpha, cyclooxygenase 2, and inflammatory arthritis. *Proc Natl Acad Sci USA* **101**: 2011–2016.
68. Quartier P., Bustamante J., Sanal O. *et al.* (2004) Clinical, immunologic and genetic analysis of 29 patients with autosomal recessive hyper-IgM syndrome due to activation-induced cytidine deaminase deficiency. *Clin Immunol* **110**: 22–29.
69. Quong M.W., Romanow W.J., Murre C. (2002) E protein function in lymphocyte development. *Annu Rev Immunol* **20**: 301–322.
70. Rada C., Williams G.T., Nilsen H. *et al.* (2002) Immunoglobulin isotype switching is inhibited and somatic hypermutation perturbed in UNG-deficient mice. *Curr Biol* **12**: 1748–1755.
71. Radic M.Z., Mascelli M.A., Erikson J. *et al.* (1989) Structural patterns in anti-DNA antibodies from MRL/lpr mice. *Cold Spring Harb Symp Quant Biol* **54**: 933–946.
72. Ravetch J.V., Bolland S. (2001) IgG Fc receptors. *Annu Rev Immunol* **19**: 275–290.
73. Revy P., Muto T., Levy Y. *et al.* (2000) Activation-induced cytidine deaminase (AID) deficiency causes the autosomal recessive form of the Hyper-IgM syndrome (HIGM2). *Cell* **102**: 565–575.
74. Robey I.F., Peterson M., Horwitz M.S. *et al.* (2004) Terminal deoxynucleotidyltransferase deficiency decreases autoimmune disease in diabetes-prone nonobese diabetic mice and lupus-prone MRL-Fas(lpr) mice. *J Immunol* **172**: 4624–4629.
75. Rogerson B.J., Harris D.P., Swain S.L. *et al.* (2003) Germinal center B cells in Peyer's patches of aged mice exhibit a normal activation phenotype and highly mutated IgM genes. *Mech Ageing Dev* **124**: 155–165.

76. Sadighi Akha A.A., Miller R.A. (2005) Signal transduction in the aging immune system. *Curr Opin Immunol* **17**: 486–491.
77. Sambhara S., McElhaney J.E. (2009) Immunosenescence and influenza vaccine efficacy. *Curr Top Microbiol Immunol* **333**: 413–429.
78. Sayegh C.E., Quong M.W., Agata Y. *et al.* (2003) E-proteins directly regulate expression of activation-induced deaminase in mature B cells. *Nat Immunol* **4**: 586–593.
79. Schlissel M., Voronova A., Baltimore D. (1991) Helix-loop-helix transcription factor E47 activates germ-line immunoglobulin heavy-chain gene transcription and rearrangement in a pre-T-cell line. *Genes Dev* **5**: 1367–1376.
80. Schoetz U., Cervelli M., Wang Y.D. *et al.* (2006) E2A expression stimulates Ig hypermutation. *J Immunol* **177**: 395–400.
81. Sharma S., Dominguez A.L., Lustgarten J. (2006) High accumulation of T regulatory cells prevents the activation of immune responses in aged animals. *J Immunol* **177**: 8348–8355.
82. Shi Y., Yamazaki T., Okubo Y. *et al.* (2005) Regulation of aged humoral immune defense against pneumococcal bacteria by IgM memory B cell. *J Immunol* **175**: 3262–3267.
83. Shlomchik M., Mascelli M., Shan H. *et al.* (1990) Anti-DNA antibodies from autoimmune mice arise by clonal expansion and somatic mutation. *J Exp Med* **171**: 265–292.
84. Shlomchik M.J., Marshak-Rothstein A., Wolfowicz C.B. *et al.* (1987) The role of clonal selection and somatic mutation in autoimmunity. *Nature* **328**: 805–811.
85. Sigvardsson M., O'Riordan M., Grosschedl R. (1997) EBF and E47 collaborate to induce expression of the endogenous immunoglobulin surrogate light chain genes. *Immunity* **7**: 25–36.
86. Sims G.P., Shiono H., Willcox N. *et al.* (2001) Somatic hypermutation and selection of B cells in thymic germinal centers responding to acetylcholine receptor in myasthenia gravis. *J Immunol* **167**: 1935–1944.
87. Sokol R.J., Booker D.J., Stamps R. *et al.* (1998) Autoimmune hemolytic anemia caused by warm-reacting IgM-class antibodies. *Immunohematology* **14**: 53–58.
88. Steger M.M., Maczek C., Berger P. *et al.* (1996) Vaccination against tetanus in the elderly: do recommended vaccination strategies give sufficient protection? *Lancet* **348**: 762.
89. Sylvestre D.L., Ravetch J.V. (1994) Fc receptors initiate the Arthus reaction: redefining the inflammatory cascade. *Science* **265**: 1095–1098.
90. Ta V.T., Nagaoka H., Catalan N. *et al.* (2003) AID mutant analyses indicate requirement for class-switch-specific cofactors. *Nat Immunol* **4**: 843–848.
91. Takahashi S., Nose M., Sasaki J. *et al.* (1991) IgG3 production in MRL/lpr mice is responsible for development of lupus nephritis. *J Immunol* **147**: 515–519.
92. Takai T., Ono M., Hikida M. *et al.* (1996) Augmented humoral and anaphylactic responses in Fc gamma RII-deficient mice. *Nature* **379**: 346–349.
93. Taylor G.A., Carballo E., Lee D.M. *et al.* (1996) A pathogenetic role for TNF alpha in the syndrome of cachexia, arthritis, and autoimmunity resulting from tristetraprolin (TTP) deficiency. *Immunity* **4**: 445–454.

94. Tchen C.R., Brook M., Saklatvala J. *et al.* (2004) The stability of tristetraprolin mRNA is regulated by mitogen-activated protein kinase p38 and by tristetraprolin itself. *J Biol Chem* **279**: 32393–32400.
95. Teng G., Hakimpour P., Landgraf P. *et al.* (2008) MicroRNA-155 is a negative regulator of activation-induced cytidine deaminase. *Immunity* **28**: 621–629.
96. Theofilopoulos A.N., Dixon F.J. (1985) Murine models of systemic lupus erythematosus. *Adv Immunol* **37**: 269–390.
97. Tran T.H., Nakata M., Suzuki K. *et al.* (2010) B cell-specific and stimulation-responsive enhancers derepress Aicda by overcoming the effects of silencers. *Nat Immunol* **11**: 148–154.
98. van Der Put E., Frasca D., King A.M. *et al.* (2004) Decreased E47 in senescent B cell precursors is stage specific and regulated posttranslationally by protein turnover. *J Immunol* **173**: 818–827.
99. van Es J.H., Gmelig Meyling F.H., van De Akker W.R. *et al.* (1991) Somatic mutations in the variable regions of a human IgG anti-double-stranded DNA autoantibody suggest a role for antigen in the induction of systemic lupus erythematosus. *J Exp Med* **173**: 461–470.
100. Weksler M.E., Szabo P. (2000) The effect of age on the B-cell repertoire. *J Clin Immunol* **20**: 240–249.
101. Wellmann U., Letz M., Herrmann M. *et al.* (2005) The evolution of human anti-double-stranded DNA autoantibodies. *Proc Natl Acad Sci USA* **102**: 9258–9263.
102. William J., Euler C., Christensen S. *et al.* (2002) Evolution of autoantibody responses via somatic hypermutation outside of germinal centers. *Science* **297**: 2066–2070.
103. Winkler T.H., Fehr H., Kalden J.R. (1992) Analysis of immunoglobulin variable region genes from human IgG anti-DNA hybridomas. *Eur J Immunol* **22**: 1719–1728.
104. Yang X., Stedra J., Cerny J. (1996) Relative contribution of T and B cells to hypermutation and selection of the antibody repertoire in germinal centers of aged mice. *J Exp Med* **183**: 959–970.

Index